The Web of Life

Perspectives in Science and Engineering
A series of anthologies which explores the relationship of science and engineering with human civilisation.
Edited by S.K. Biswas, Indian Institute of Science and C.V. Vishveshwara, Indian Institute of Astrophysics, Bangalore, India.

Volume 1
The Web of Life
Edited by G. Padmanaban, M. Biswas, M.S. Shaila and S. Vishveshwara

Other Volumes in Preparation
Engineering as a Human Enterprise
R. Narashima and S.K. Biswas

The Web of Life

edited by

G. Padmanaban

M. Biswas

M.S. Shaila

and

S. Vishveshwara

Indian Institute of Science
Bangalore, India

harwood academic publishers

Australia • Canada • China • France • Germany • India • Japan • Luxembourg
Malaysia • The Netherlands • Russia • Singapore • Switzerland • Thailand
United Kingdom

Printed in Singapore.

Amsteldijk 166
1st Floor
1079 LH Amsterdam
The Netherlands

British Cataloguing in Publication Data

The web of life. - (Perspectives in science and engineering)
1. Biology
I. Padmanaban, G.
574

ISBN 3-7186-5927-1

CONTENTS

PREFACE

Biology today is one of the most exciting and challenging branches of science. For a long time it was an empirical discipline devoted primarily to the observation and classification of living organisms. Over the years, systematic studies in biology have paved the way to an understanding of inter-relationships within the living world. Such understanding crystallised into conceptual and theoretical frameworks such as evolution and genetics. These in turn placed biology within a time frame and established its nexus to a broader universal setting. Biology has now grown into a rich and mature science involving interactions among its sub-branches as well as with other areas of the sciences. One of the most recent and dramatic developments in biology is its ability to alter and control the structure and function of living beings at the genetic level. This has raised profound moral and ethical questions.

The articles in this volume cover a wide spectrum of inter-related topics. Some of them consider in some detail developments in specific areas such as molecular biology, immunology and evolution, while others address broader issues. A major theme that emerges is based on interactions at different levels. One of interaction is interdisciplinary. Mathematics has played a continually evolving role in providing a framework for describing biology. Similarly, thermodynamics has been a powerful tool in understanding biological processes. Another level of interaction is among different species. Even the study of a simple virus exemplifies the unity in nature. Interactions within a given species are reflected in inbreeding in human societies. At the level of ethics and morality, biology has introduced in recent times an altogether new debate. Today, social accountability of biologists, shared responsibility in preserving biodiversity, the consequences of and necessary precautions in influencing natural processes are issues engendered by biology, which have profound implications for human survival.

The essays in the present volume written for a wide audience offer a glimpse into the realms of modern biology. They trace the all-encompasssing web that has life as its nucleus and touches many spheres of human endeavor. They form, we believe, a mosaic that portrays the present day revolution in biology that will be carried well into the future.

We are grateful to all the contributors who not only responded warmly to our invitation to write but also co-operated with us during a rather long period of preparation for this volume. We thank Harwood Academic Publishers for bringing out this volume.

We acknowledge the valuable help and advice of Professor Vidya Nanjundiah of the Indian Institute of Science at different stages of the project. Finally we thank Ms. N. Pushpa for secretarial assistance and Mr. D. Satyan for his help with the preparation of some of the figures.

G. Padmanaban
Margaret Biswas
M.S. Shaila
Saraswathi Vishveshwara

CONTRIBUTORS

Alan H. Bittles
Edith Cowan University
School of Applied Science
Perth
Western Australia 6027
Australia

Baruch S. Blumberg
Fox Chase Cancer Centre
7701, Burholme Avenue
Philadelphia
Pennsylvania 19111
USA

Graham R. Flannery
School of Genetics and Human Variation
La Trobe University
Bundoora
Victoria 3083
Australia

Alan L. Mackay
Department of Crystallography
Birkbeck College
University of London
Malet Street
London WC1E 7HX
UK

Philippe Marliere
Department des Biotechnologies
Institut Pasteur
25, rue du Dr. Roux
75015 Paris
France

Rupert Mutzel
Department of Biology
University of Konstanz
Postfach 5560
D-7750 Konstanz 1
Germany

G. Padmanaban
Indian Institute of Science
Bangalore, 560 012
India

Peter A. Parsons
Department of Genetics and Human Variation
La Trobe University
Bundoora, Victoria 3083
Australia

Anita Rattan
Musvaagevej 6, 2 tv
DK-8210 Aarhus-V
Denmark

Suresh I.S. Rattan
Laboratory of Cellular Ageing
Department of Chemistry
Aarhus University
DK-8000 Aarhus-C
Denmark

N.H. Ravindranath
Centre for Ecological Sciences
Indian Institute of Science
Bangalore, 560 012
India

Gordon A. Rodley
Centre for Peace and Conflict Studies
The University of Sydney
Mills Building A26
NSW 2006
Australia

M.S. Shaila
Indian Institute of Science
Bangalore, 560 012
India

1. Interaction of Species

Baruch S. Blumberg

Fox Chase Cancer Centre Philadelphia, Pennsylvania, U.S.A.

INTRODUCTION

The theme of this book is the web of life, an inquiry into how the elements of our natural world are interconnected, and the consequences of this. Seeking unity and connection in nature has been a goal of science since the realization that analytical science has great explanatory powers. One of the major goals of 17th, 18th and 19th century science was to collect vast amounts of data, arrange these in museums and encyclopedias, and from this derive unifying generalizations. This goal of unity now appears to have a lower priority in biology, probably because modern science amasses such an enormous amount of data, that simplification seems beyond our immediate grasp.

However, another approach to demonstrating the unity of the elements of the natural world is to start from the other end. That is, to select a miniscule organism and determine how it relates to other members of the great family of living things, and in so doing, learn about their relationships; start from the particular and proceed to the general.

The mechanism I will use for this maze-exploration will be the hepatitis B virus (HBV), an infectious agent which despite its apparent simplicity (a 420 Angstrom diameter and a genome of only 3200 base pairs it is among the smallest of the human viruses), has managed to establish a secure niche for itself in a complicated environment and has infested a very large percentage of the present and past inhabitants of the world. We will try to place ourselves in the mind-set of the virus, if an organism with only four reading frames and very little in the way of what we recognize as ordered perception and communicative capabilities, can be said to have a mind. In this essay, I will give examples of how our study of HBV has led to the investigation of the biology, behaviour and even the psychology of humans and other organisms that share the environment of the virus.

HBV AND ITS INTERACTION WITH HUMANS

Adaptability, will, intent, are terms usually ascribed to the behaviours of higher organisms with complex nervous systems. It would be curious to attribute these qualities to a virus but the following anecdotes might raise these possibilities, if only in a humorous sense.

It appears that HBV has adapted itself to the computer era to increase its transmission possibilities. Several members of the staff, patients and attendants at a busy hospital developed hepatitis in a pattern that suggested linkage between them. An investigation of the outbreak showed that computer cards had been used to carry information between the laboratory and other parts of the hospital. Computer cards have sharp edges; these occasionally inflict minute but deep cuts through which HBV can be transmitted when the blood of a patient infected with HBV comes in contact with the cards. The cards become carriers of infection, since replicating HBV can survive for a period outside its host. The epidemic apparently was spread in this manner and the problem disappeared when this was recognised.

In a second example, HBV managed to intrude itself into a complex game activity developed by humans. Orienteering is a demanding sport popular in Scandinavia and elsewhere. The contestants are equipped with a map, compass and a set of instructions for a route to a series of geographic stations to which they must report. The winners are those who reach all the stations in a specified or in the shortest time. In the wooded landscape of northern countries this often requires running through uncleared brush and undergrowth. There was an epidemic of hepatitis B, which eventually affected several hundred Orienteerers. An investigation of the epidemiology revealed that the runners usually wore only shorts. Their lower legs were lacerated by the brush and, from time to time, they would stop at rest stations, clean their legs with water, and then dry themselves with any available towel. The same towel, and apparently, the same water, were used by different runners. If even a single person were infected, then many people could be subsequently exposed to HBV and infected in turn. This resulted in the epidemic, which was easily remedied when the sequence of events was understood and the runners were required to wear long trousers.

These stories illustrate how the incredibly small HBV organism, in fulfilling its strategies for spread and replication, interacts with the complex behaviours of humans. As we shall see, in subtle and not-so-subtle ways, it can also change human behaviours.

HUMAN BEHAVIOUR AND THE TRANSMISSION OF HBV

Hepatitis B virus is spread by the transmission of blood from an infected person to one who is not infected. What are the human behaviours by which blood can be transmitted? There are several curious human practices, which can actually or theoretically transmit viruses in blood. We examined the ethnographic literature and museum collections to identify some of these.

In many folk circumcision ceremonies, several people are circumcised using the same instrument at about the same time. In folk practice, the instrument would not be sterilized between procedures. HBV is very hardy and cannot be killed without autoclaving or extended exposure to boiling water; even harsh disinfectants may not render the vaccine non-infectious. HBV carried in the blood of one individual involved in the circumcision practice could be therefore transmitted to many others.

Tattooing is widely practiced in many cultures. Traditionally, the same instrument might be used by a folk practitioner to decorate many people without adequate cleaning or sterilization between uses. Tattooing is often used to signify that the tattooed individuals belong to a family, tribe, clan or other social grouping. It has been shown that this procedure can transmit HBV readily from one individual to another. Hence, ironically, members of the social group will have their affiliation demonstrated not only by the externally obvious tattoo but also by sharing an infection, an unwanted indicator of social bonding. Surgical instruments used for blood letting, still a common form of treatment in many parts of the world, may transmit blood from one individual to another. A common form of skin decoration, in some regions, is scarification of the skin surfaces in a pleasing design. Since the scarification instrument is not sterilized between uses, it can serve as a vehicle for transmitting the virus between individuals in the same social group.

There are many methods of human-to-human transmission of blood that are all too common in contemporary "developed" societies. A major mechanism for HBV transmission results from injection of illicit drugs. Needles are often shared and, in many cases, a group of users may use the same syringe in a session of drug use. There is little interest in sterilizing their equipment or in using disposable equipment, hence, the virus is very readily transmitted. Another curious observation has been that drug abusers worldwide may share the same subtype of HBV, as if the world community of drug abusers were in effect "blood brothers", identified by the specific HBV that has been transmitted from blood stream to blood stream through the medium of the drug use.

Identification of HBV Carriers and the Problems of Stigmatization

Viruses can cause disease, suffering and misery by their direct biological effects. They can also have indirect effects which are equally detrimental as a consequence of their interactions with human perceptions. When, in 1967, it became possible to identify carriers of HBV, there was a rapid and enthusiastic acceptance of the technique for clinical diagnosis and for the screening of transfusion donor blood. Within a few years, millions of people were being tested yearly for the presence of the virus. Soon after the introduction of carrier testing, we were consulted by individuals who had been identified as carriers and who had been seriously affected by this information, so much so that we feared we were witnessing the development of a stigmatized class. Little was known at the time about the infectious propensities of carriers. In some locations, employees of health care organizations who were identified as carriers were being asked to find other employment. There were suggestions in the medical literature that nurses should be tested after graduation and those found to be HBV carriers prohibited from practicing their profession. In some cases, surgeons and dentists found to be carriers were not allowed to practice. In one country with a high frequency of carriers, it was questioned whether all those found to be carriers should be excluded from military officer training and from medical schools. A question was raised as to whether children from regions with high frequencies of carriers should be screened for HBV before being presented as candidates for adoption. The child's fate might be determined by the results of a single blood test for an occult character, which might not otherwise have been perceived by the carrier or the potential adoptive parents.

All this was happening despite the lack of knowledge of how infectious carriers might be. There was a conflict between the possible health benefits to the general public and the personal rights of the carriers. There was a clear advantage to society of the screening of blood donors; it decreased the amount of post-transfusion hepatitis. It was not a great disadvantage to the donor who was asked not to donate blood in the future. However, there was no advantage to general screening of populations, since nothing could be done for the carrier, and at that time little of practical value could be arranged for the individuals with whom the carriers might come into contact. It was obvious that more research was required to determine what to do when carriers were identified.

The ethics of this problem were widely discussed. The general consensus among workers in the HBV field was to encourage blood donor screening but not screening of the general population. The emphasis was to be placed on encouraging basic research to develop a fuller understanding of HBV which, in turn, would lead to a better knowledge of how to proceed with application. Within the next few years, significant gains in knowledge led to a partial reversal of the ethical position. Methods for distinguishing highly infectious carriers from those of lower infectivity were developed. The vaccine which we had invented in 1969 was developed and by the early 1980s became widely available. The nature of the relation between HBV and primary cancer of the liver became better understood. Techniques for medical surveillance of carriers and detection of cancer at an early stage were institutionalized. Research on developing an antiviral agent that would rid the carrier of the virus or decrease its pathological effect intensified. All these research developments, in effect, changed the ethical situation, and now in many cases it became of advantage to both the individual carrier and society to screen for carriers. This represents an interesting example of how changes brought about by research have reversed the ethical position created by earlier research.

HEPATITIS B AND RELATED VIRUSES IN ANIMALS AND BIRDS

HBV was originally found in humans. Shortly after its discovery, a serological survey was made of a large number of terrestrial, avian, and marine species. Using the available insensitive method of immunodiffusion in agar gel, the virus was detected in several non-human primates, but not in other species. The infection was most common in the higher apes, gorillas, chimpanzees, orangutan and others, and also, less commonly, in other primate species including African green monkeys, macaques, and spider monkeys. In some of the higher apes the infection was quite common; for example, many of the chimpanzees used in the early "space monkey" flights were infected. The apes, in most of the cases, did not appear to become ill, but they were infectious carriers of the virus and often would transmit the disease to their human handlers and others with whom they came into contact.

Few studies have been conducted in primates in the wild state, but when they were, infection was not found. It appeared likely, therefore, that the animals had become infected after capture by contact with the humans who trapped, caged and cared for them; but the primates, in turn, could be a source of infection for other humans with whom they interacted. This could be of epidemiological importance in areas, such as parts of India and Africa, where the patterns of behaviour of human and non-human primates bring them in close proximity to each other. An example of this could be the troops of primates which are frequently found in temples and in the densely packed market-places of the tropics. HBV, on the face of it, appears to be an infectious organism which, in its present form, is a pathogenic organism in humans, but apparently not in other primates.

It is likely, however, that in the remote past there has been a richer interaction between animals (and possibly plants) and viruses which are related to HBV, HBV-like viruses have now been identified in several mammals and birds. In some, for example, the woodchuck, tree squirrel, ground squirrel, three banded palm tree squirrel, duck and blue heron, the similarity to HBV is remarkable. Woodchuck hepatitis virus (WHV) has about 80% DNA homology with human HBV, striking cross reactivity with the surface and core antigens, and similar DNA polymerases. The disease they cause in the different species, chronic liver disease and primary cancer of the liver, are often remarkably similar in appearance and in pathogenesis. A hierarchy of relations between these viruses can be developed, based on the degree of homology, the methods of spread and the nature of the disease patterns.

The several animal and avian HBV-like viruses provide a web of unity between their host species whose study can reveal an interesting level of interdependence in nature.

VIRUSES SIMILAR TO HBV WHICH INFECT PLANTS

A group of plant viruses, The Cauliflower Mosaic Viruses, have similarities to HBV and HBV-like viruses, including significant base pair homologies for certain sections of the genome. There were also several reports (which have not been confirmed) that tobacco plants may be infected with HBV under certain circumstances. These observations raise the intriguing possibility that plant substances effective in protecting plants from viral infection may also be useful as mammalian antiviral agents.

There are other interesting implications which derive from the long and intimate inter-relations which have existed over countless generations between animals and plants. It can be inferred that the biochemical, immunological and other polymorphism of plants and animals have interacted and accommodated to each other. Most medicines we currently use are derived ultimately from natural substances, and hints as to their functions in plants may provide some indication of their potential uses as therapies in humans. Most of these applications have been as a consequence of empiric experience and a surprising number of the medicines we now use were first used in indigenous medical systems.

Based on these and other considerations, we attempted to identify possible antiviral agents for HBV. We determined which genera of plant had been used for the treatment of jaundice (one of the most characteristic symptoms of infection with HBV) in several indigenous medical systems. Based on this approach, we have investigated a medication derived from *Phyllanthus amarus* a plant widely used in India, Africa, South America, Asia, Oceania and elsewhere, for the treatment of jaundice and other symptoms thought to be associated with liver disease. The investigation of the derivatives of this plant and other possible treatments are in progress.

Subtypes of HBV and Human Variation

There is a complex pattern of subtypes of HBV. The relative frequency of these subtypes can vary from population to population in a manner which appears to be related to the genetic composition of the population. For example, there are significant differences in the distribution of HBV subtypes between French- and English-speaking Canadians. In Japan, inhabitants of the Ryuku Islands stretching to the south of the main islands have subtype distributions which are more similar to the Malaysian and Indonesian people to the south, while inhabitants of the more northern islands have affinities with the subtypes of HBV found in Korea and north Asia. Intermediate frequencies can be found in the other islands, presumably reflecting the patterns determined by the ancient peopling of Japan. These subtypes have a pattern of ordered distribution similar to that of genetic polymorphism such as the red cell blood types, the HLA antigens and the RFLP's (restriction enzyme fragment length polymorphism) which are used for the genetic characterization of individuals and populations. From this it can be inferred that there are inherited characteristics in humans which determine their ability to become chronically infected with a particular subtype of HBV. This implies the existence of polymorphic genes in humans that are related to HBV and which are present before the individual has become conventionally infected by the virus. There are, in fact, a series of polymorphic sequences at multiple locations in the human genome which have homologies with the X gene of HBV. The X gene may be responsible for the integration and the initiation of replication of HBV, and the X-related genome locations in humans may relate to the sites at which HBV can integrate.

This raises one of the most interesting areas for conjecture about viruses and their hosts. Where do viruses come from? Do they derive from the genomes of other species, breaking off at some point in the process of cell division and assuming an independent "life" of their own, but one intimately connected with their species of origination?

The corollary of this question is that of the contribution that viruses might make to the genome of higher species. Retroviruses, HBV and other viruses can integrate into the genome of their host. How large a contribution can viral genes make to the host genome? How often do these sequences enter the zygote cells and thereby contribute to the genetic makeup of the generations following that of the individual initially infected? Are these the sequences that control the ability to become chronically infected with HBV and insuring its perpetuation through many generations? Further, it is interesting to question if these genes may contribute to the evolution of species in that their integration could, under some circumstances, have the functional equivalence of a "mutation" on which selection could occur.

Infection with HBV and the Sex Ratio at Birth in Humans

The proportion of males and females in a population is of biological, economic, cultural and psychological importance. The rate of growth of the population is dependent on the number and proportion of females of childbearing age, and the social structure of a population is very sensitive to sex ratio. Curiously, there is a significant body of data linking the human host response to infection with HBV, and the ratio of males to females at birth. This data can be interpreted to mean that infection with HBV has a major impact on sex ratio and therefore on the basic character of human populations.

In several populations in which HBV infection is common, we found that when those infected become carriers of HBV (that is, remain persistently infected for many years) they are likely to have a higher proportion of males among their offspring than parents (particularly mothers) who develop antibody against the surface antigen of the virus (anti-HBs). These observations were consistent in populations from Greece, Papua New Guinea, Greenland and the Philippines. In some studies, the difference was a consequence of a smaller number of females in the carrier families resulting in a higher sex ratio and smaller family size. We have suggested a model that, as a characteristic of the chromosomal differences between males and females, HBV can replicate more readily in females than in males, and therefore, females would be more likely to die as a consequence of infection from their mothers or, indirectly, from their carrier fathers. We hope that additional experimental and observational data can be accumulated to develop and test this model.

There are some very interesting consequences if the sex ratio effect of HBV can be validated. Areas with high sex ratios, for example, South China, also have some of the highest frequencies of HBV carriers, as predicted from the observations and models. The explanation could be biological and due to the operation of the HBV sex ratio effect just described. Family size has been restricted in some regions where HBV carriers are common. If the parents are carriers they would, by inference from our model, be more likely to have male children. Non-carriers would have a lower probability for males and, in a traditional society, might desire to have additional children until the pregnancy resulted in a male. Widespread HBV vaccination programs should, in the long run, decrease the number of carriers. Would this have an effect on sex ratio? There is insufficient data to know if this would occur, and it would be valuable to make the appropriate observations as the vaccination programs become widespread in China and elsewhere.

CONCLUSIONS

HBV has remarkable connections with humans and other organisms and events in its environment. A study of these connections can help us to understand the inter-relations of our complex biological and social world.

Acknowledgements

This work was supported by USPHS grants CA-40737 and CA-06927 from the National Institutes of Health and by an appropriation from the Commonwealth of Pennsylvania.

2. A Study in Human Population Genetics: The Influence of Marriage Patterns on the Gene Pool

Alan H. Bittles

Edith Cowan University, Perth and King's College, University of London, U.K.

INTRODUCTION

During the course of the last decade, much attention has focused on rapid technological developments in molecular genetics which collectively have facilitated a major increase in our understanding of the nature of the human genome. At the same time there has been a relative lack of progress in the field of human population genetics, although it is widely accepted that the extrapolation of knowledge from the individual to the population level is essential, if we are to fully utilize the information which has been gathered from investigations into human genome structure and function. A specific example of human population genetics is examined in this essay, namely, the types of sexual unions which are undertaken by humans, with particular emphasis placed upon the outcome of marriages between close biological relatives. In keeping with the overall aims of the volume this topic has been approached from a multidisciplinary angle, with the intention that by so doing a more complete and realistic appreciation of the total subject will result.

THE HISTORICAL BACKGROUND

One of the more curious aspects of the development of human society is the manner in which certain populations choose to marry close biological relatives, whereas others strenuously avoid this type of marital union. The predominant Western stereotype of inbreeding is of a small, poor and remote community, in which a large proportion of the inhabitants suffer from obscure physical disorders and exhibit obvious symptoms of mental sub-normality. Implicit in this thinking is the belief that all the ills of such communities are the direct result of matings between closely related individuals. While in most instances this type of characterization is almost certainly a myth, geographically isolated, inbred communities occasionally have been documented with, in a number of cases, a high prevalence of physical and/or mental defects. However, what has seldom if ever been demonstrated is whether or not these defects are actually associated with the expression of deleterious recessive genes or instead, that they result primarily from disadvantageous environmental circumstances, dietary iodine deficiency being a well-known example.

To a large extent, the generally negative views on close kin marriage which still prevail in Western society may have been fostered by, if not resulted from, historical religious and secular prejudices. According to the Venerable Bede in his text *Ecclesiastical History of the English People*, written in the 8th century AD, on appointment as the first Archbishop of Canterbury in AD 597, the Primate Augustine wrote to Pope Gregory I seeking clarification on the position of the Latin Church with regard to unions between close kin. In reply the Pope confirmed a ban on marriages up to and including third cousins and, although the prohibition on second and third cousin unions was formally rescinded in 1917, to the present day specific dispensation remains a prerequisite for Roman Catholic first cousins who wish to have their marriages solemnized in Church. While the Christian Orthodox Churches similarly proscribe consanguineous unions, the reformed Protestant denominations have no such rules, citing as their guide-lines scripture contained in the Biblical Book of *Leviticus*. Thus, within this latter tradition, first cousins are free to marry should they so choose.

A similar lack of unanimity is seen when secular regulations on marriage are examined within and between countries. Under civil statutes introduced in the U.S.A. from the mid-19th century onward, first cousin marriages are criminal offences in eight of the fifty states of the Union and are subject to civil sanction in a further thirty-one legislatures. Yet, in the state of Rhode Island, those of the Jewish faith are permitted uncle-niece (but not aunt-nephew) marriages. By comparison, in the U.K. there are no civil laws to limit or to discourage marriages between first cousins or more distantly related individuals, whatever their religious persuasion or personal inclination.

Despite prevailing adverse opinions with respect to consanguineous unions, i.e., marriages between couples "of the same blood", during the mid- to late 19th century many eminent persons were married to close relatives. A prominent example was the marriage of Charles Darwin and his first cousin Emma Wedgwood, a granddaughter of Josiah Wedgwood the founder of the world famous pottery. Darwin appears to have been genuinely concerned by the widespread Victorian belief that their offspring might be biologically disadvantaged due to the effects of inbreeding, even though seven of the ten children survived to adulthood with several attaining positions of academic eminence. For this reason, he persuaded Sir John Lubbock to petition the British Government for the inclusion of a question on the prevalence of cousin marriage in the 1871 Census of Great Britain and Ireland. Unfortunately for Darwin, the proposal was dismissed by a Parliamentary Select Committee on the grounds that it could lead to embarrassment and distress both among couples who had contracted cousin marriages, and their children. Darwin's vigorous response to this Parliamentary rebuttal subsequently appeared in the book *The Descent of Man*, where he stated with obvious annoyance: "When the principles of breeding and of inheritance are better understood, we shall not hear ignorant members of our legislature rejecting with scorn a plan for ascertaining by an easy method whether or not consanguineous marriages are injurious to man".

Historical sources of this nature are of considerable anecdotal interest, and they clearly demonstrate that avoidance of consanguineous unions was by no means universal in 19th century Western society. Collectively however, their limited experimental design, with little or no control for social, economic or demographic variables, severely limits any potential application to assessment of the effects of inbreeding in contemporary human populations, and by extension the influence that inbreeding may have had on the human genome through generational time. In the present essay this topic will be examined under a series of sub-headings, including the current global prevalence of consanguineous unions, social and economic correlates of consanguineous unions, the effects of consanguinity on reproductive behaviour, premature mortality associated with inbreeding, and the reported effects of inbreeding on physical and mental morbidity.

THE GLOBAL PREVALENCE OF CONSANGUINEOUS MARRIAGE

Even in countries or communities where marriages between close relatives formerly were widespread, it has been widely assumed that their prevalence has greatly diminished in recent generations. This supposition certainly holds true for North America and Western Europe, with marriage rates at first cousin level in the general population of 0.6% or less, and to a lesser extent in Japan, where the national prevalence of first cousin unions has reduced from an estimated 5.9% to 8.0% in the early 1950s to 3.9% in the 1980s. Yet consanguineous marriages remain strongly favoured in many other regions. This is illustrated in Tables 1 and 2, which are based on ninety-eight studies variously conducted at local, state and national levels within the last four decades.

TABLE 1. Prevalence of Consanguineous Marriages by Region and Country

Less than 1%	1–10%	20–50%	*Unknown*
North America	East Africa	North Africa	Middle Africa
Europe	West Africa	Central Asia	South Africa
Russia	South America	South Asia	Caribbean
Oceania	North India	West Asia	Central America
	Kazakhstan		East Asia
	Japan		South–East Asia

Source: Bittles (1990)

As Table 1 shows, in the mainly Muslim countries of North Africa, Central and West Asia, and in many parts of South Asia, marriages contracted between persons who are related as second cousins or closer account for 20% to over 50% of the observed total in the present generation. In 1994 the combined population of these regions was estimated to be 732 million, and a further 1468 million persons live in countries where 1% to 10% of marriages are contracted between biological relatives

(Table 2). It should be stressed that these estimates are intentionally conservative in nature, as they ignore both the 41% of the world's population for which representative and reliable consanguinity data are currently unavailable, and the presence of large numbers of highly inbred immigrant families in many Western countries. Yet, even after adopting this cautious approach, it is apparent that consanguineous marriage is still the preferred form of marital union in many of the world's major human populations.

TABLE 2. Estimated Global Numbers of Consanguineous Progeny

Less than 1%	10008 *million*
1–10%	1468
20–50+%	732
Unknown	2399
Global population	5607

Sources: Bittles (1990); Population Reference Bureau (1994)

Considerable variation has been reported between populations in the specific types of inbred marriages which are contracted. For example, the so-called parallel first cousin form of marriage predominates in many Muslim societies, where a man marries his father's brother's daughter. In terms of population genetics this results in a coefficient of inbreeding (F) for their children of 0.0625. That is, the offspring are predicted to have inherited identical gene copies from each parent at 6.25% of gene loci, over and above the baseline level of homozygosity in the general population. Less frequently, double first cousin marriages also are reported between Muslim couples ($F = 0.125$), where the spouses have both sets of grandparents in common. By comparison, although on average 20% to 45% of marriages in the primarily Hindu states of South India are contracted between close relatives, the most popular forms of consanguineous union are uncle-niece ($F = 0.125$) and cross first cousin, usually between a man and his mother's brother's daughter. As an aside, it is interesting to note that while double first cousin and uncle-niece unions represent the same genetic distance, the latter form of marriage is prohibited by the Koran and so in most Muslim societies it is virtually unknown. In South and West Asian countries the followers of many other religions, including Buddhists, Christians, Jews, Parsees and Druze, also frequently choose to marry close biological kin. While this is generally taken to be an indication of their acceptance of prevailing local marriage norms, for some small religious minorities living in remote areas there may also be the practical issue of very restricted marriage partner choice, effectively leading either to marriage between individuals who are biological relatives or celibacy.

Information on the prevalence of consanguineous marriage tends to be limited in many other major populations. Anthropological and ethnographic surveys have recorded a preference for, or at least acceptance of, cousin marriage in approximately 35% to 50% of populations in the sub-Saharan region of Africa, which probably reflects both traditional societal practices and the gradual influence of Islamicization. China is a country of special interest given its population of 1200 million, and the fact that little or no information appears to have been collected on consanguinity since the establishment of the People's Republic. Prior to the Second World War, the mother's brother's daughter form of first cousin marriage was an accepted custom among the Han who make up some 90% of the total population, and marriage with a close relative also was widely contracted in several of the minority populations, notably the Muslim Uighur of Xinjiang province in the west of China. An indication that consanguinity has persisted among the Han comes from reports commencing in 1990, which refer to proposed strict governmental limitations on marriages between first or second cousins on "eugenic" grounds. Clearly, this is a somewhat contentious subject and one that merits close future attention. There also is a notable shortage of data on consanguineous marriage from Indonesia, which in population terms is the world's fourth largest country with an estimated population of over 200 million. Whether this lack of information should be interpreted as indicating the avoidance of such marriages is uncertain, although anthropological sources do suggest that at least in the south-eastern islands of Indonesia, cross-cousin unions were permissible in former generations.

As examples of the types of inbreeding currently practised in South Asia, information on consanguineous marriages typically contracted in South India and in the Pakistan province of Punjab is reproduced in Figures 1 and 2. The South Indian data were collected from women delivering their babies in hospital in the cities of Bangalore and Mysore, Karnataka, whereas the Pakistani results were obtained via a series of urban household surveys. In general, unions contracted between persons related as second cousins or closer are categorized as consanguineous. This limit is chosen because the genetic influence in marriages between couples related to a lesser degree would be expected to differ only marginally from that observed in non-consanguineous unions.

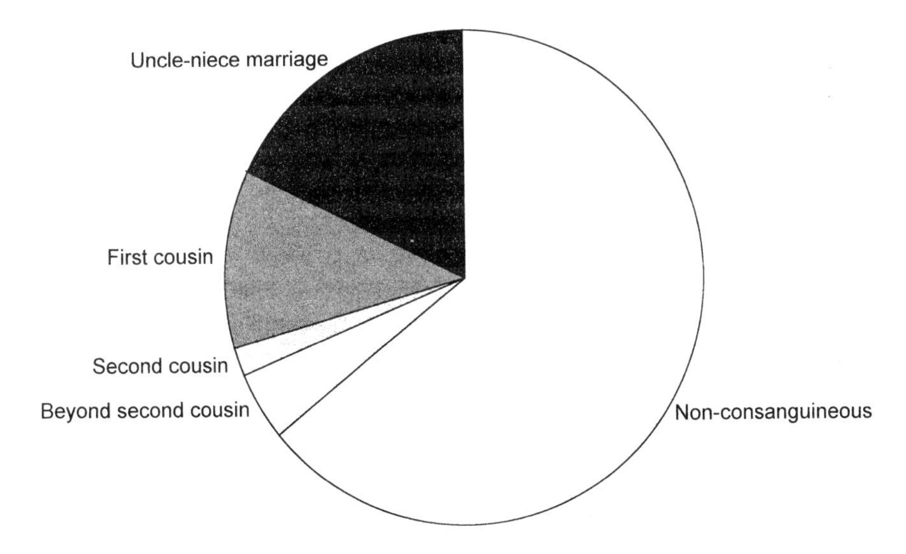

Figure 1. Distribution of consanguineous marriages in Karnataka, South India, 1980–1989. Uncle–niece ($F=0.125$); first cousins ($F=0.0625$); second cousins ($F=0.0156$); beyond second cousins ($F<0.0156$); non-consanguineous ($F=0$).

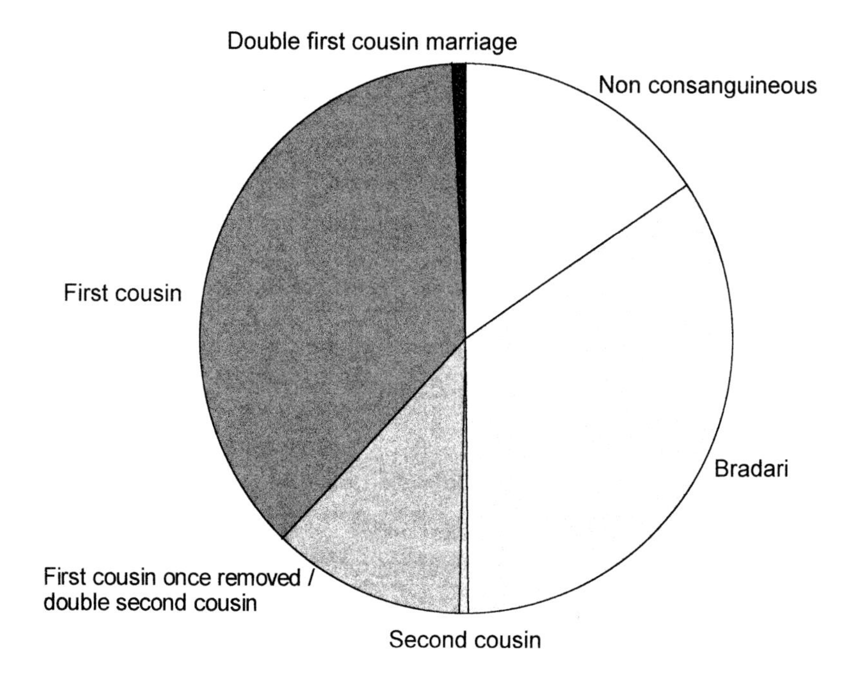

Figure 2. Distribution of consanguineous marriages in Punjab, Pakistan, 1979–1985. Double first cousins ($F = 0.125$); first cousins ($F = 0.0625$); first cousins once removed/double second cousins ($F = 0.0313$); second cousins ($F = 0.0156$); *bradari* unions ($F < 0.0156$); non-consanguineous ($F = 0$).

It can be seen that the actual patterns of marriage in the two regions differ according to local custom. Thus in South India where 31% of marital unions are consanguineous, the most common forms of marriage are between uncle and niece (especially in the majority Hindu community) and first cousins (Figure 1). By comparison, in Pakistan the survey method revealed a wider variety of marriage forms, including double first cousin (equivalent to a coefficient of inbreeding in the progeny of ($F = 0.125$), first cousin ($F = 0.0625$), first cousin once removed and double second cousin ($F = 0.0313$), second cousin ($F = 0.0156$), *bradari* ($F < 0.0156$) and non-consanguineous unions ($F = 0$). Over 50% of marriages are contracted between spouses related as second cousins or closer, and a further 34% of marriages are between spouses drawn from the same *bradari* (literally translated as brotherhood). In *bradari* marriages the partners are believed to be more distantly related than second cousins ($F < 0.0156$), although because of the imprecise nature of the relationships calculation of an F value is impossible.

Since the consanguinity classifications refer to marriages in the present generation only, the average F values calculated for each locality represent minimal consanguinity estimates. Despite the much higher percentage of consanguineous unions in Pakistan than South India, because of the South Indian preference for uncle – niece unions the average F values (indicating the intensity of inbreeding in each population) are actually quite similar. Thus for Pakistan, $F = 0.0280$ and for South India, $F = 0.0299$. The F values for both populations indicate that their overall levels of inbreeding in the current generation approximate to that of first cousins once removed.

When assessing the mean levels of genetic identity in a population two additional factors require consideration. First, all endogamous societies with breeding pools of finite size, whether small, isolated communities or societies in which inflexible social sub-divisions exist, will exhibit some degree of random inbreeding because of restricted marriage partner choice. Irrespective of positive or negative attitudes to consanguinity within the community, this form of population sub-division effectively results in under-estimation of the average coefficient of inbreeding for the population. Second, in societies with a long tradition of consanguinity the cumulative value for F would be expected to greatly exceed that calculated for a single generation, once again resulting in under-estimation of the true level of homozygosity in the population gene pool.

SOCIAL AND ECONOMIC CORRELATES OF CONSANGUINITY

Among populations in which consanguinity is preferential, the highest rates of marriage to a close relative are consistently reported in more traditional rural areas and among the poorest and least educated sections of society, although in Japan and South India major land-owning families may provide a partial exception to these generalizations. In all cases the maintenance of family property is a major consideration, with only token or significantly reduced dowry or bride-wealth payments required when marriages between close biological relatives are contracted. Among land-owning families however, the preservation of their estates and land-holdings is probably the critical factor in choosing to marry close kin. Besides these economic considerations, consanguineous marriages are preferred because of the comparative ease with which the traditional marriage negotiations can be conducted. It is further believed that marriage between close relatives offers the optimum opportunity for social compatibility, both between husband and wife and the bride and her mother-in-law. In turn, this greater social compatibility is thought to considerably benefit female status within the community as a whole. Perhaps of greatest importance, there is a strong underlying conviction that, by marrying within the extended family, hidden uncertainties regarding health or other unfavourable family characteristics will not arise as a source of future disagreement.

Two separate and additional considerations specifically apply within Islamic society. The first is the belief that in contracting a father's brother's daughter (first cousin) marriage, the partners benefit from the close comparability in status of their respective fathers. Secondly, there is the precedent that two of the wives chosen by the Prophet Muhammad were biological relatives, one of whom was his first cousin. The Prophet also married his daughter Fatima to Ali, who was both his ward and the son of his paternal uncle. Therefore by marrying within the family, devout Muslims could be said to be emulating the example set by the Prophet.

Interactions between consanguinity and a wide range of social variables may complicate assessment of the genetic effects of human inbreeding to a significant degree. As previously noted, in most communities consanguineous marriages are more likely to be arranged among the poorest and least educated families. Since it frequently has been observed that the offspring of these families suffer disproportionately from a high prevalence of infectious and nutritional diseases, any failure to control for socio-economic differentials could readily lead to exaggerated estimates of the adverse effects of inbreeding. On the other hand, in those countries where consanguineous marriages are widely practised, the omission of parental inbreeding as a contributory factor in health studies throws into question the exact nature of the roles played in postnatal survival by a number of widely accepted sociodemographic factors. Typically these are parameters such as maternal age, maternal education, birth order, and birth interval.

CONSANGUINITY AND REPRODUCTIVE BEHAVIOUR

Both biological and social considerations must be taken into account when assessing the relationship of consanguinity to reproductive behaviour. From a biological viewpoint, the sharing of common human leucocyte antigens (HLA) by spouses, a situation which by definition would be more probable in consanguineous marriages, has been claimed to be a contributory factor among couples experiencing difficulties initiating pregnancies, and/or with a history of recurrent spontaneous abortion. The precise mechanism(s) governing recurrent abortion remains obscure, and there is preliminary evidence that the problem may not be due to the HLA antigens themselves but rather be caused by HLA-linked genetic defects which interfere with normal development of the embryo. This remains a poorly understood and somewhat controversial topic, particularly as the collection of data on early spontaneous abortion has been shown to be hampered by substantial recall problems.

In fact, surveys conducted in major, inbred populations have consistently shown reduced levels of primary sterility in consanguineous marriages, usually interpreted as stemming from the greater immunological compatibility of the mother and fetus and thus reducing the risks from potentially lethal conditions such as rhesus incompatibility and pre-eclamptic toxaemia. There also has been little or no evidence linking inbreeding to increased rates of spontaneous abortion or miscarriage, and similarly there is no indication that multiple birth rates or the sex ratio at birth are subject to inbreeding effects. Therefore unless deleterious recessive genes are operational very early in pregnancy, in effect before the female's first missed menstrual period, consanguinity does not appear to adversely influence the incidence of prenatal losses at population level to any significant degree.

Many studies have shown that greater numbers of infants are born to closely related spouses than to non-consanguineous couples. A recent comparative survey used data collected in North and South India, and Pakistan to assess the extent of this trend, and in seventeen of the twenty populations examined a positive association between consanguinity and fertility was confirmed. Besides the biological factors referred to above, a number of social factors may be strongly implicated in the greater fertility of consanguineous marriages. Younger maternal age at marriage is particularly important in this respect, since it facilitates increased levels of fertility by optimizing maternal reproductive span and concentrating child-bearing in the most fertile years. This pattern was clearly seen in the South Indian cities of Bangalore and Mysore, where, in terms of age at first delivery, mothers in uncle-niece unions were 1.4 years younger, and first and second cousins 1.1 years and 0.8 years younger respectively than women in comparable non-consanguineous unions. In all three cases the net result was greater numbers of live births in each consanguinity category than in the non-consanguineous control group. A second factor of this nature which requires consideration is the tendency for mothers in consanguineous unions to continue child-bearing at older ages. Once again this effective prolongation of fertility would be expected to lead to the birth of greater numbers of children to consanguineous couples. However it also could result in the birth of more children with chromosomal anomalies, an example being Down syndrome (trisomy 21), since a clear relationship between advanced maternal age and the risk of a child being conceived with this disorder has long been appreciated.

It has been suggested that the greater number of births in consanguineous marriages may also represent some form of reproductive compensation response to increased early postnatal mortality. This could operate either via a conscious decision by parents to achieve their desired family size, and/or the cessation of lactational amenorrhoea following the death of a breast-fed infant. Unfortunately, on the basis of information so far available from South Asia, it has not been possible to determine whether the greater fertility of consanguineous unions also is a facet of reproductive compensation. Although in Pakistan larger numbers of live births have been reported in families where childhood deaths have occurred, especially when the deceased child was male.

CONSANGUINITY AND PREMATURE MORTALITY

In overall biological terms, the net effect of inbreeding is to increase the numbers of homozygotes and, at the same time, to reduce the numbers of heterozygotes in the breeding pool. The adverse health effects specifically associated with consanguinity are caused by the expression in homozygotes of rare, recessive genes inherited from a common ancestor, and in populations where inbreeding is common, increased levels of morbidity and mortality caused by the action of detrimental recessive genes can be predicted. As indicated in the introduction, many impressions and opinions as to the effects of inbreeding have stemmed from studies conducted on geographical, religious and ethnic isolates. Examples of each of these categories are the island of Tristan da Cunha in the South Atlantic and Newfoundland in Canada, the Samaritans in Israel and the Amish in Pennsylvania, U.S.A., and Travellers in Ireland and Gypsies in the U.S.A. Even in the absence of preferential consanguinity, alleles which are rare in large populations can randomly increase to high frequency in small groups within a few generations, because of the twin phenomena of founder effect and genetic drift in a breeding pool of restricted size.

To examine the effects of inbreeding on deaths prior to puberty, a study was conducted in eleven towns and cities of the Pakistani province of Punjab. As expected, total pre- and postnatal mortality uncorrected for possible socio-economic differentials was found to be significantly associated with degree of inbreeding. As a percentage of all reported pregnancies, mortality increased from 16.4% in non-consanguineous progeny, to 20.1% in second cousins, 22.9% in first cousins once removed and double second cousins, 22.1% in first cousins and 39.0% in double first cousins (Figure 3). The mortality differentials across all five consanguinity classes were least obvious in reported abortions /miscarriages, and they exerted proportionately greatest effect during the first year of life, indicating that congenital defects, and hence the expression of detrimental recessive genes, may have been a significant factor determining the excess mortality associated with inbreeding. Because of the absence of control for socio-economic variation the results must be treated with caution, in particular the very high death rates among double first cousin progeny during the first month of life when young maternal age and associated gynaecological immaturity may have been a significant contributory factor. From an epidemiological perspective, these data suggest that in Pakistan 41% per cent of all pre-reproductive deaths in children born to double first cousins were associated with the expression of detrimental recessive genes, with equivalent values for first cousins of 27%, and 15% and 8% in first cousins once removed/double second cousins and second cousins respectively. Calculation of the fraction of total mortality associated with consanguinity in the study sample indicated that 18% of all late prenatal and childhood deaths in Punjab were associated with inbreeding.

To see whether these excess death rates applied universally, a further survey based on data from thirty-eight populations in India, Pakistan, Japan, Kuwait, Nigeria,

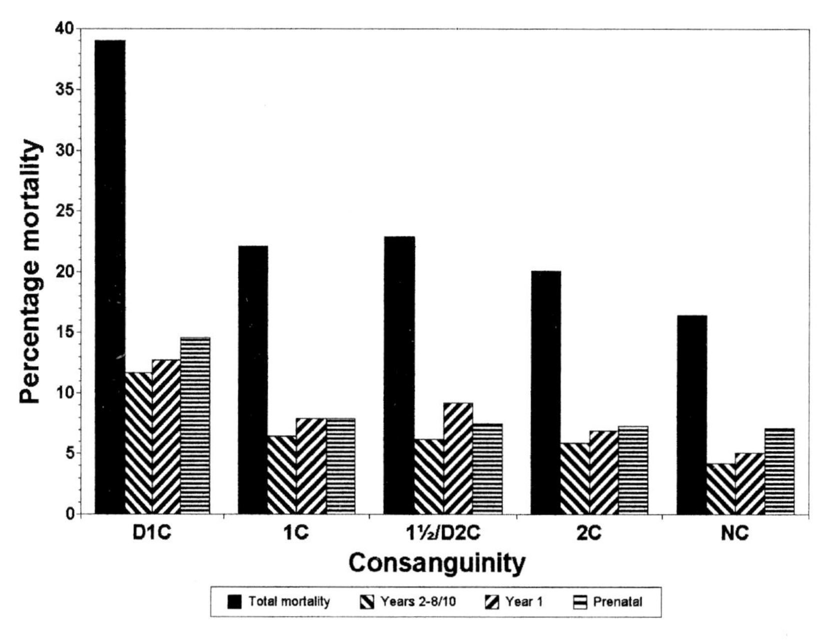

Figure 3. Percentage mortality by consanguinity category at differing ages in Punjab, Pakistan, 1979–1985. D1C, double first cousins; 1C, first cousins; 1.5CD2C, first cousins once removed/double second cousins; 2C, second cousins; NC, non-consanguineous.

Brazil and France was organized. The information derived from this large transnational study, based on 622,188 pregnancies and live births, indicated that mean excess mortality among the progeny of first cousins was 4.4% higher than in non-consanguineous offspring. On the basis of these findings it could be further calculated that mortality was 1% to 2% higher among the progeny of second cousins and 8% to 10% higher in those of uncle-niece and double first cousin unions. The data can also be used to derive an estimate of the number of lethal equivalents in the human genome, i.e., the number of alleles which when homozygous would lead to the death of an individual. The figure that emerges from this study is 0.7 lethal equivalents per gamete, or 1.4 lethal equivalents per zygote, which is significantly lower than had been calculated in earlier investigations employing much smaller sample sizes.

CONSANGUINITY AND MORBIDITY

Failure to control for socio-economic variables also may be an important source of extraneous variation in studies into the effects of inbreeding on ill-health. For example, a large survey in Japan showed that if socio-economic differentials between consanguineous and non-consanguineous families had not been controlled, the calculated effect of inbreeding on various body measurements would have been inflated by approximately 20%. When social variables have been subject to control, measures with a polygenic multifactorial pattern of inheritance appear to be little affected by consanguinity. Thus studies into the influence of inbreeding on body measurements at birth and in childhood have failed to reveal any major, consistent pattern, and only marginal declines were shown in the mean scores attained by consanguineous progeny in tests of intellectual capacity. In the latter case, it would appear that inbreeding mainly leads to greater variance in IQ levels, due in part to the expression of detrimental recessive genes in a small proportion of those tested.

The incidence of major congenital malformations generally has been reported to be higher in consanguineous than in non-consanguineous progeny. However, since most communities which favour consanguineous marriage would be categorized as economically less developed, there is continual difficulty in ensuring both an unambiguous diagnosis of the disorder and adequate discrimination between genetic and non-genetic determinants of morbidity. It has been possible to overcome these problems in studies on the Pakistani community resident in the U.K., among whom approximately 55% of marriages are arranged between first cousins. Mean perinatal mortality in the Pakistani community at 15.7 per thousand significantly exceeds that in the indigenous population (7.0 per thousand) and in other ethnic groups (where it averages 9.6 per thousand), and this is a pattern which appears to be consistent across all socio-economic classes. Multi-ethnic prospective studies in the British Midlands also indicate that serious malformations occur in 28.2 per thousand of Pakistani babies, chronic diseases/disabilities are diagnosed in 41.5 per thousand of those who survive the first month of life, and congenital anomalies account for 41% of all deaths within the first year. A preliminary survey which was conducted in Pakistan further suggested that a number of the major diseases of adulthood, including some forms of cardiovascular disease and cancers, may be more prevalent in inbred progeny.

DISCUSSION

Considered on a global basis, it is apparent that consanguineous marriage is the preferred or even in some cases the prescribed choice for a significant proportion of the world's large population groups, especially those resident in many regions of Asia and Africa. Frequently it has been claimed that a rapid reduction in the prevalence of consanguineous unions would inevitably occur during the latter part of the present century, with industrialization, greater population movement, a decline in family size, and higher literacy rates cited as the main factors contributing to this change.

There are a number of reasons for believing that predictions of this nature are incorrect. First, they ignore the fact that in South India and Pakistan, and many other parts of the world, consanguinity is not merely a cultural tradition but is judged to offer significant social and economic benefits. During the last two decades these benefits have become ever more sharply focused in South Asia, partially as a result of the increasing expectations and demands for exorbitant dowry payments, which have caused severe financial hardship within families and been linked to the controversy of "dowry deaths" in a number of the North Indian states. Under these circumstances, and where culturally permissible, marriage to a close relative is generally regarded as a more sure and economically feasible choice than contracting a marital union with a non-relative. Second, with major improvements in public health facilities in many of the world's less developed regions, including greater provision of vaccination for the prevention of maternal tetanus following child-birth and for childhood infectious diseases, larger numbers of children will survive to marriageable age. In turn, it would be expected that a greater proportion of the community would choose to exercise their preference for marriage to a close biological relative, a prevalence which previously may have been constrained by a lack of suitable partners. Finally, increased rates of marriage to a close relative can be envisaged in those parts of the Islamic world where fundamentalist doctrines have been gaining favour, accompanied by a return to more traditional practices.

In the short term, the only major factor that can be identified as significantly impeding this projected increase in consanguineous unions would be the introduction of proscriptive Governmental legislation, as recently reported in China. It is however unclear just how effective such legislation would be in most democratic countries. For example, although uncle – niece marriages were prohibited in India by the Hindu Marriage Act of 1955, between 1980 and 1989 they nevertheless accounted for 21% of all Hindu unions sampled in the South Indian state of Karnataka. As evidenced by the results of the Karnataka survey, conducted in the technologically advanced cities of Bangalore and Mysore even in regions where a decline in consanguineous unions does take place, the pace of change will be slow and it will principally occur across generations. For South Asia, this prediction would particularly apply to rural areas in which literacy is lowest, resistance to modernization is strongest, and a large majority of the population are resident. (Currently an estimated 52% of the population of India are illiterate and 74% live in rural areas. In Pakistan the comparable figures are 65% illiteracy and 72% of the population are rural inhabitants.)

TABLE 3. Population Projections for More and Less Developed Countries

Year	1994	2010	2025
More developed	1164	1228	1259 million
Less developed	4443	5794	7119 million
World total	5607	7022	8378 million

Source: Population Reference Bureau (1994)

As shown in Table 3, high population growth rates are projected for the less developed countries, with in many cases the highest rates predicted for regions where consanguineous marriages are widely contracted. On balance therefore, an increase in the total numbers of inbred progeny appears probable in the immediate future. Perhaps less obviously, a similar change may be in train at local level in several countries in Western Europe, North America and Australasia which have experienced substantial migration from Asia, the Middle East and Africa during the last four decades. Furthermore, any move towards the imposition of tighter immigration controls by these Western countries would presumably only serve as an additional positive factor in encouraging consanguineous unions among established minority ethnic communities with a restricted pool of potentially acceptable marriage partners.

Switching to the longer-term view, it would be expected that through time the improved survival prospects of children in economically less developed countries will lead to a decrease in completed family sizes. If this assumption proves correct, then future generations may gradually begin to experience difficulty in obtaining close biological kin to marry, thus leading to a decline in the prevalence of consanguineous unions and/or to some relaxation in the preferred types of consanguineous marriages. The latter factor may explain an apparent accommodation between the traditional preference for mother's brother's daughter cross-cousin marriage and partner availability in the metropolitan Hindu population of Madras, Tamil Nadu, with reports that parallel cousin marriages have been increasingly contracted during the course of the last decade. Intuitively, it will be the prevalence of the biologically closest forms of marriage which will be affected to the greatest extent by any such change, leading in particular to declines in the prevalence of uncle – niece and double first cousin unions.

Survey results from a wide variety of sources show that fertility is greater in consanguineous unions. This greater fertility may act as a contributory factor in the higher postnatal mortality observed among consanguineous progeny, since the expectation of having an affected child increases with increasing family size. But, as noted earlier, all studies conducted in less developed regions face a major problem in diagnosing cause of death and, under such circumstances, it is difficult to partition mortality into genetic and non-genetic components with any real degree of confidence.

For the moment, evidence that excess mortality among consanguineous progeny in countries such as India and Pakistan has a significant genetic component is largely dependent on statistical analysis, which is clearly unsatisfactory. Studies from South India, where a high proportion of childhood morbidity was specifically diagnosed as genetic in origin, and in the U.K. – resident Pakistani community, provide strong support for the hypothesis that deleterious recessive genes significantly determine childhood morbidity and mortality in communities with a high prevalence of consanguineous unions. It is however essential that the actual levels of expressed genetic defect in these communities be kept in perspective, and also that the outcome of consanguineous marriages is not subject to assessment simply in terms of comparative medical audit.

A further, major consideration is the manner in which consanguinity acts as a covariable with other sociodemographic parameters. Numerous surveys have emphasized the importance of maternal education as a determinant of childhood mortality. When data from these sources have been examined, it is apparent that the major decline in mortality during the last four decades has been among deaths in the 1–4 age group, as might be expected where vaccination programmes for common childhood infectious diseases have been successfully implemented. The levels of success achieved in reducing infant deaths unfortunately have been much less encouraging, particularly during the first month of life. It therefore seems probable that in countries where marriage to a close biological relative is preferential, a significant proportion of early postnatal mortality reflects the effect(s) of inbreeding and hence the expression of deleterious recessive genes, rather than being determined by sociodemographic factors alone. Again, this is a topic which to date has received little attention, largely because of the paucity of suitable data.

As socio-economic conditions improve and the incidence of primarily environmental disease concomitantly declines, genetic disorders will account for an increasing proportion of global morbidity and death. This pattern has already been observed in the more developed, low mortality countries such as the U.K. and, as has been shown above with respect to the U.K. – resident Pakistani community, it is especially obvious where consanguineous marriages are commonplace. Based on experience to date in the more developed Western countries, it would be appropriate if planning for this major transition in the disease profiles of many of the more populous developing countries is rapidly instituted. Otherwise, in future years the task of providing life-time care for individuals with chronic inherited disease states may prove to be not only difficult, but also be regarded as a major financial and emotional burden. Of course, given a reduction in the popularity of consanguineous unions, fewer homozygotes would be conceived leading to reduced numbers of persons adversely affected with recessive disorders.

The final topic to be considered in this review on the theme of population genetics concerns the nature of the relationship between consanguinity and the composition of the gene pool. There has been a popular theoretical belief that, in communities where inbreeding has been conducted at high frequency over multiple generations, genes which are detrimental to health would have been selectively eliminated from the gene pool because of the reduced viability of persons homozygous for such alleles. Viewed positively, if somewhat insensitively, this theory suggests that the practice of consanguineous marriage would have been to the long-term good of the species, albeit at the expense of certain severely disadvantaged individuals. While this altruistic aspect of consanguinity may have its attractions, as yet there is no convincing experimental evidence from human populations to support the theory. It should also be noted that the greater number of surviving progeny observed in the families of consanguineous spouses makes any such "cleansing of the gene pool" inherently improbable, since for most common recessive disorders selection against heterozygotes is believed to be very low.

What remains is an intriguing natural experiment in human population genetics that has been conducted in many communities over thousands of generations, in effect since the dawn of human evolution and the earliest development of human societies in their many and varied forms. Perhaps via the application of newly available DNA probing techniques, definitive conclusions as to both the positive and negative effects of consanguinity on the gene pool may soon be within our grasp. Until such time as this information does become available, further governmental attempts to legally restrict consanguineous unions should best be placed on hold, and speculation as to their desirability or otherwise which lacks solid experimental foundations equally would best be avoided. In any event, past experience would suggest that attempts to coerce communities into changing their traditional marriage patterns are unlikely to meet with either approval or success.

Acknowledgements

The valued collaboration of my colleagues Professor N. Appaji Rao and Dr. H.S. Savithri, Indian Institute of Science, Bangalore and Dr. Sajjad A. Shami, Quaid-i-Azam University, Islamabad is gratefully acknowledged. The work was made possible by generous financial assistance provided by the Wellcome Trust.

Glossary

Alleles: alternative forms of a gene at a specific locus.
Chromosome: the basic genetic structural unit along which genes are arranged.
Coefficient of inbreeding: a statistical measure of the proportion of gene loci at which an individual is homozygous. Can also be used to describe the mean level of inbreeding in a population.
Consanguineous: mating between individuals who share at least one common ancestor. Conventionally applied to persons related as second cousins or closer.
Endogamy (*endogamous*): the practice of marrying within one's local community.
Founder effect: the current frequency of a gene in a population which can be traced back to one of the founders.
Gamete: the male and female genetic components of a zygote.
Gene: a defined sequence of DNA which codes for a particular product, e.g., a specific protein.
Heterozygote (*heterozygous*): an individual who has inherited a normal copy of a gene from one parent and an abnormal copy of the same gene from the other parent.

Homozygote (*homozygous/autozygous*): an individual who has inherited identical copies of a gene from each parent.
Human leucocyte antigens (*HLA*): genetic markers found on the surface of most nucleated cells of the body, e.g., the leucocytes. Important in transplantation reactions and in providing defence against microbial attack.
Lethal equivalent gene: a gene which results in the prereproductive death of an individual. Alternatively, two alleles at different loci, each of which when homozygous results in 50% mortality prior to the age of reproduction, or four alleles, each of which when homozygous causes 25% prereproductive mortality, etc.
Locus: the specific position of a gene on a chromosome.
Polygenic multifactorial disorder: a disease resulting from the interaction of multiple genes in combination with unfavourable environmental factors.
Random genetic drift: the chance fluctuation of gene frequencies in a small population.
Recessive: a disorder which is only expressed in homozygous individuals.
Zygote: the fertilized ovum, formed from the fusion of a sperm and an egg.

Further Reading

Monographs

Bittles, A.H. 1990. *Consanguineous Marriage: Current Global Incidence and its Relevance to Demographic Research. Population Studies Center, Research Report Number 90–186. Ann Arbor: University of Michigan.*

Bittles, A.H. and Roberts, D.F. (eds.) 1992. Minority Populations: Genetics, Demography and Health. London: Macmillan

Articles

Bittles, A.H., Mason, W.M., Greene, J. and Appaji Rao, N. 1991. Reproductive behavior and health in consanguineous marriages. *Science* **252**: 789–794.

Bittles, A.H., Grant, J.C. and Shami, S.A. 1993. An evaluation of consanguinity as a determinant of reproductive behaviour and mortality in Pakistan. *International Journal of Epidemiology* **22**: 463–467.

Bittles, A.H. 1994. The role and significance of consanguinity as a demographic variable. *Population and Development Review* **20**: 561–584.

Bittles, A.H. and Neel, J.V. 1994. The costs of human inbreeding and their implications for variations at the DNA level. *Nature Genetics* **8**: 117–121.

Basic Demographic Data

World Population Data Sheet 1994. Population Reference Bureau Washington, D.C.

3. Evolutionary Change: A Phenomenon of Stressful Environments

P.A. Parsons

Division of Science and Technology, Griffith University, Nathan, Brisbane, Queensland 4111, Australia.

INTRODUCTION: IS STRESS UNIVERSAL?

In 1859, Darwin published *On the Origin of Species by Means of Natural Selection, or the Preservation of Favoured Races in the Struggle for Life*. For a workable theory, Darwin considered how one species evolves into a different species in terms of individual organisms and populations. In current terminology, selection acts upon the organisms that make up populations in a given generation. The mechanism of selection follows the observation that individuals differentially adapted to the environment reproduce at varying rates. Darwin put forward the view that in any environment an organism will accumulate over time the variations best fitting it to its surroundings. As the environment changes, new variants will become advantageous, and will tend to supplant variants that have become less well adapted. He appreciated that the effectiveness of this process of natural selection depends upon the necessity that the variations must be inherited and available for times of environmental change. This organism-environment interaction is central to our understanding of evolution today.

The publication of *The Origin of Species* followed the simultaneous presentation to the Linnean Society of London, the theory advanced by Darwin and Wallace (1859) that species are not individually created and unchanging, but that species can evolve into new species during the course of time. Although Darwin observed and meticulously recorded variation in both domestic and wild plants and animals, he did not understand the mechanism underlying the inheritance of variation. Indeed, following these initial publications came the period of Nineteenth-Century Darwinism which was primarily concerned with the naturalist's observations of the physical world, and with applications and debates concerning the concepts of evolution as put forward at this time. The rediscovery of Mendelian heredity in 1900 led to the

Correspondence: PO Box 906, Unley, SA 5061, Australia.

rise of modern genetics, and ultimately the interpretation of the variation observed by Darwin in terms of genetic principles. Notable contributors to this synthesis were R.A. Fisher, J.B.S. Haldane and Sewall Wright. Fisher's (1930) *The Genetical Theory of Natural Selection* provides a succinct account of this phase which developed into the synthetic theory of evolution. This phase was largely completed by 1949 (Mayr, 1982a). These writings led to an understanding of genetical variation of organisms in some cases relatable to their environments, a process which continues to this day. However in 1953, Watson and Crick explained the structure and function of DNA, which was followed by a rapid increase in the understanding of the mechanisms of heredity and variation of living organisms. As each new molecular development occurred, the changes underlying evolutionary change became more precisely documented. The enormous and ever expanding literature on the analysis of protein variation by electrophoresis, restriction enzyme analysis, the study of transposable elements and the technique of DNA finger-printing, all testify to this ferment.

How do these remarkable developments of the molecular biologist relate to organism-environment interactions? Enormous bodies of data have been accumulated on enzyme variation following electrophoresis for discussions concerning the relative importance of various evolutionary forces underlying change at the protein level. Nevo, Beiles and Ben-Shlomo (1984) retrospectively assessed much of this information covering a wide range of taxa, and found that the major force producing evolutionary change at the protein level is at the level of ecological parameters, many of which incorporate environmental extremes. Indeed, based upon the wide array of habitat types that were involved, temperature itself emerged as the variable of underlying significance, which is understandable given its importance in all biological processes. Even though such an *a posteriori* procedure is very inefficient, it does highlight the importance of temperature extremes. A more specific example comes from a survey of genetical variation in Californian populations of the slender wild oat, *Avena barbata,* where microgeographical variation measured from an assessment of electrophoretic variants is only interpretable in terms of adaptation to variation in temperature extremes and aridity regimes among habitats (Clegg and Allard, 1982).

As studies at the molecular level become increasingly sophisticated, there is a parallel need that molecular variation in natural populations should be studied incorporating ecological considerations in an *a priori* sense. The study of organism-environment interactions is as relevant today as in 1859. The only difference is that the type of genetic variation under study alters as our understanding of this variation increases. Since extremes have been invoked when relating molecular studies to the environment, it is appropriate to ask whether Darwin was aware of the possible evolutionary importance of environmental extremes. Perusal of the *Origin of Species* reveals a number of references to extreme external conditions, principally climate:

> "Climate plays an important role in determining the average numbers of a species, and periodical seasons of extreme cold or drought seem to be the most effective of all checks."
> "When we reach the Arctic region, or snow-capped summits, or absolute deserts, the struggle for life is almost exclusively with the elements."
> "But a plant on the edge of a desert is said to struggle for life against the drought."

and in Darwin and Wallace (1859), the joint publication of the theory of evolution, Darwin wrote:

> "It should be remembered, that in most cases the checks are recurrent yearly in a small, regular degree, and in an extreme degree during unusually cold, hot, dry, or wet years, according to the contribution of the being in question."

Even so, Darwin regarded evolution to be mainly a gradual process emphasizing intricate and complex interactions among species, and to a lesser extent, between species and their physical environment.

As Mayr (1982a) points out, A.R. Wallace based his conclusions on rather strictly ecological arguments. His major claim to fame following the publications on the theory of evolution was his *Geographical Distribution of Animals* (1876), the classic of zoogeography for many decades. Two quotes are indicative of his approach:

> "Climate appears to limit the range of many animals, though there is some reason to believe that in many cases it is not the climate itself so much as the change of vegetation consequent upon climate that produces the effect."
> "Any slight change, therefore, of physical geography or of climate, which allows allied species hitherto inhabiting distinct areas to come into contact, will often lead to the extermination of one of them; and this extermination will be effected by no external force, by no actual enemy, but merely because the one is slightly better adapted to live, to increase, and to maintain itself under adverse circumstances, than the other."

His emphasis upon climate in determining the distribution of species is clear, but more in terms of changes of the mean than extremes.

While numerous books on evolution have appeared in this century, an important landmark was Schmalhausen's (1949) book, *Factors of Evolution*. In the foreword to the 1986 edition, D.B. Wake considers that this book is a direct descendent of Darwin incorporating the synthetic theory of evolution as the following indicates:

> "The focus is never on single loci nor on isolated genes in populations, but on organisms, the interactions that influence their development, and variation among them and its consequences."

The emphasis of the book is upon the organismic level of biological organization incorporating a sophisticated approach to genotype-environment interactions. Extreme environments are considered to be important, for example:

> "The continental populations are highly resistant to extremes of temperature-both high and low."

Arguing from laboratory selection experiments for cold tolerance in *Drosophila*, Schmalhausen appreciated that natural selection for resistance to environmental extremes could be effective over quite short time periods:

> "All these facts prove that the dependence between an organism and its external environment can be altered rapidly and that the responses of an organism to its external environment are historically conditioned."

In summary, phenotypic selection underlies organic evolution, so that genetic changes should occur following phenotypic changes especially when environments are stressful.

Anderwartha and Birch (1954) largely view the distribution and abundance of each species to be dependent upon the combination of the many physical and biotic factors needed for survival and reproduction at the individual level. Within the dipteran species, *Drosophila melanogaster,* ecological phenotypes relating to factors determining distribution and abundance tend to vary clinally in a manner determined by *a priori* considerations of geographic variations in climate (Stanley and Parsons, 1981). It is, in fact, periods of short-term high and low temperature stresses that are the most effective in defining climatic races of *D. melanogaster,* a result which extends to the kinetic properties of enzymes such as alcohol dehydrogenase (Alahiotis, 1982). Furthermore, as shown in Figure 1, the broad distributions of

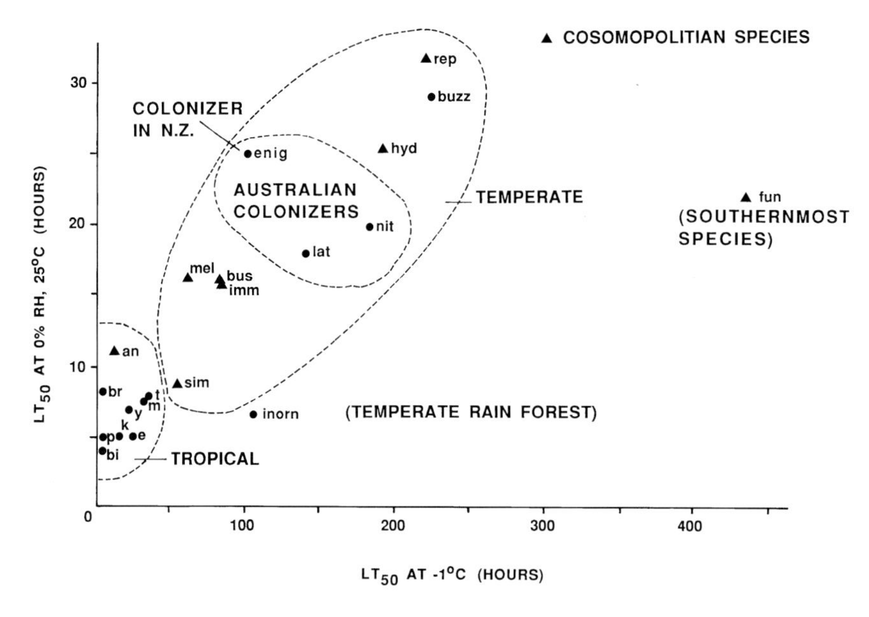

Figure 1. LT_{50} values expressed as a plot of the number of hours at which 50% of flies (sexes combined) died of −1°C stress Vs 0% RH 25°C stress for various *Drosphila* species: an, *ananassae;* bi, *bipectinata;* br, *birchii;* bus, *busckii;* buzz, *buzzatii;* e, *erecta;* enig, *enigma;* fun, *funebris;* hyd, *hydei;* imm, *immigrans;* inorn, *inornata;* k, *kikkawaii;* lat, *lativittata;*m, *mauritiana;* mel, *melanogaster;* nit, *nitidithorax;* p, *paulistorum;* rep, *repleta;* sim, *simulans;* t, *teissieri;* y, *yakuba,* (Source: Adapted from Parsons, 1981).

Australian *Drosophila* are relatable to extreme temperatures and tolerance to desiccation (Parsons, 1981). Similar emphases upon climatic extremes come from studies on the distributions of Australian floral types (Nix, 1981), which have now been formalized into the development of a bioclimatic prediction system (BIOCLIM) incorporating annual, seasonal, and extreme components of the environment (Busby, 1986a). From known distributions of species, this BIOCLIM system is then used to determine their bioclimatic envelopes based upon climatic similarities with actual distributions. One of the major aims has been to determine the distributional limits of species, and it has been successfully used in a number of contexts in Australia, including the distribution of the rainforest tree, *Nothofagus cunninghami*, kangaroos, and chromosomal taxa of the Australian grasshopper, *Caledia captiva* (F) (Busby, 1986b, Caughley *et al*., 1987; Kohlmann, Nix and Shaw, 1988). In kangaroos, three species were considered with ranges overlapping to some extent. Using the BIO-CLIM system, these species were separated by a discriminant analysis where the X-axis features annual mean temperature, maximum temperature of the hottest month, minimum temperature of the coldest month and mean annual precipitation, and the Y-axis features annual mean temperature, precipitation of the driest quarter, and the temperature of the driest quarter. Substantial components of extreme stress are represented by these axes, as is the situation for the other examples cited above.

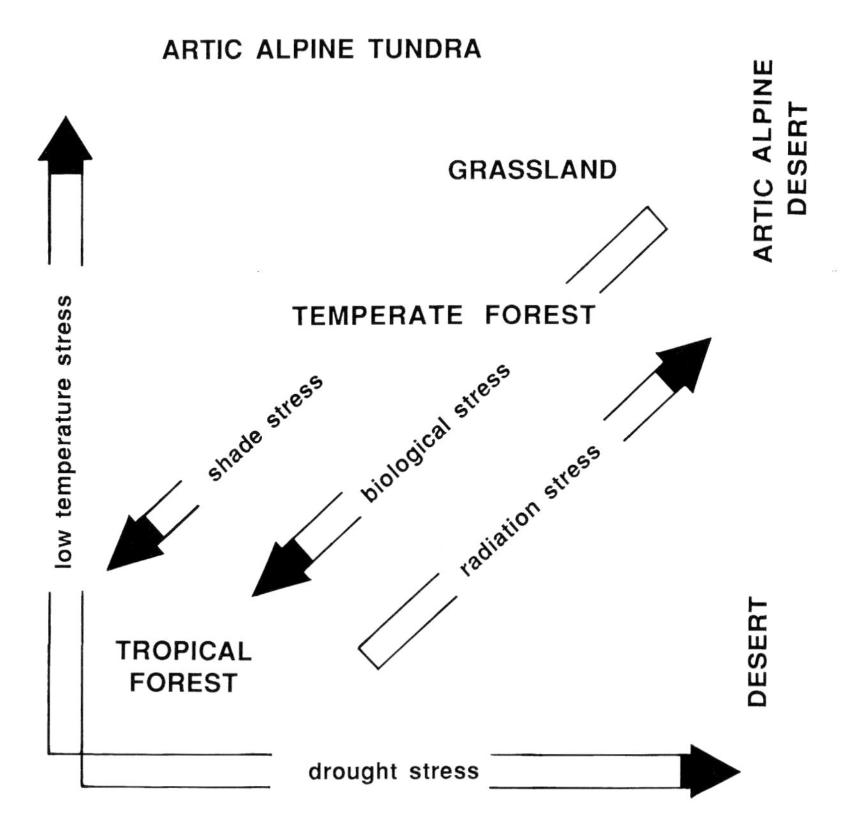

Figure 2. Schematic gradients of environmental stress in realation to major vegetational types. The X axis is a gradient of decreasing temperature, and the Y axis represents decreasing water availability (after Osmond *et al* 1987).

General discussions of plant distributions also emphasize stress physiology (Fig. 2), in particular the two components of drought stress and low temperature stress

(Osmond *et al*., 1987) indicating striking parallels with the major determinants of *Drosophila* distributions. Osmond *et al.* (1987) emphasized that stress affects most plants during part of their life cycle, and Boyer (1982) estimated that environmental stress limits U.S. agricultural productivity to 25% of its potential.

Extrapolating to *Drosophila*–a genus intimately dependent upon plants for resources–is difficult, however, the short life span of *Drosophila* in the field (Rosewell and Shorrocks, 1989) suggests that stressful conditions may often occur, and there is evidence that flies are commonly starved under field conditions (Bouletreau, 1978). In summary, it is difficult not to regard stress as approaching universality in natural populations.

STRESS AND VARIABILITY

In addition to promoting rapid phenotypic shifts, Schmalhausen (1949) realized that stress has a role in increasing genetic variability under some circumstances:

> "Increased variability in populations living under unfavourable conditions, such as extremely high or low temperatures, has been recorded."

and considering mutation he comments that:

> "the cause of mutational variability may be the numerous factors of the external environment, especially when their intensities are either unusual or extremely variable (extreme temperatures, intensive insolation, extreme limits of humidity, chemical influence of salts, unusual organic compounds in the food, etc.)"

This implies that the mutation rate is unstable in heterogeneous, changing, or unusual environments. In *Drosophila* he notes the existence of unstable lines and comments:

> "This decline in stability is explained partly by a disruption of intracellular genic or chromosomal balance by means of certain mutations."

In modern terms, the analogy is transposable elements of DNA leading to increased mutation rates under conditions of genomic and environmental stress, as discussed by McClintock (1984) who points out that these elements:

> "can restructure the genome at various levels, from small changes involving a few nucleotides, to gross modifications involving large segments of chromosomes, such as duplications, deficiencies, inversions, and other more complex organizations."

Indeed, as early as 1951, McClintock reported an elevated mutation rate from the movement of mobile genetic elements in temperature-stressed maize plants, and there is now evidence that transposition rates in maize, yeast and in *Drosophila* increase with environmental temperature and with stress at the genomic level.

It is, therefore, not surprising that recombination is affected by environmental variables. Within the first decade of genetic experiments on *D. melanogaster*, Plough (1917) found that recombination increased in chromosomes 2 and 3 especially in centromeric regions, when the temperature at which female flies developed diverged upwards or downwards from 25°C, the normal culture temperature. These and later data (e.g. Grell, 1978) indicate that recombination increases above and below normal temperatures giving a U-shaped curve (Fig. 3), and limited data from other organisms are consistent (Parsons, 1988a). Recombination increases tend to be substantial at temperatures approaching lethality, so that the precise environmental conditions that define this critical boundary are needed in each species to avoid difficulties in interpretation.

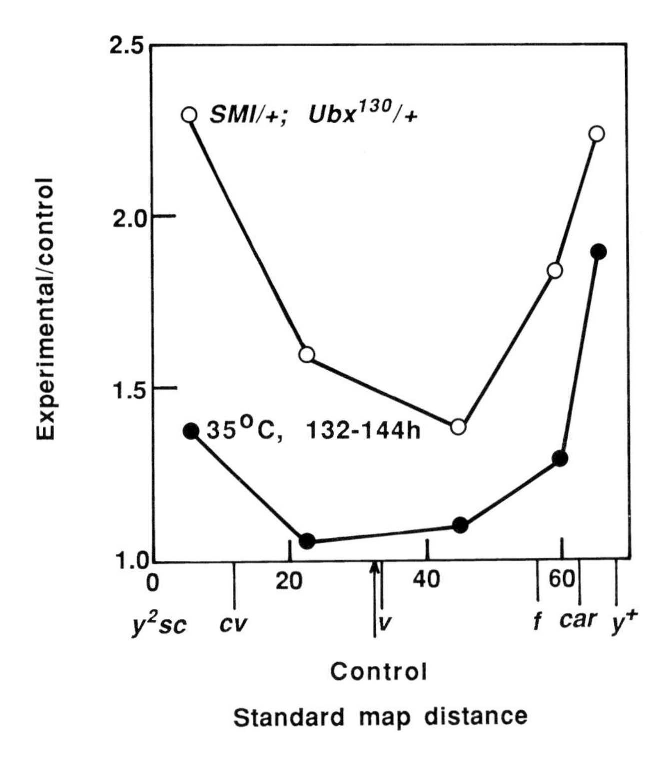

Figure 3. Crossing-over in five regions of the X chromosome of *Drosophila melanogaster* after exposure to heat stress (●) and to heterozygous inversions in chromosomes 2 and 3 (O). The graphs give the relative alterations in crossing over in the X chromosome expressed as the ratio experimental/control (after Grell, 1978).

Turning to nutrition, Bouletreau's (1978) observations on ovarian activity and reproductive potential in a natural population of *D. melanogaster* suggest the occurrence of long periods of nutritional stress in nature whereby flies only rarely have the opportunity to utilize their genetic reproductive potential. For chromosome 3 data, Neel (1941) found that recombination increased along the whole region by 0–15% following starvation.

It is important to ask whether environmental and genomic stresses are cumulative. There is little information on combinations of environmental stresses, however, there is a substantial literature on the interchromosomal control of recombination (Lucchesi and Suzuki, 1968) indicating that structural heterozygosity in one part of the genome usually increases recombination in the remainder of the genome. In parallel with the temperature extremes, the major effect is in the centromeric region (Schultz and Redfied, 1951), and there are data (Fig. 3) showing strikingly similar patterns for heat treatment and interchromosomal effects (Grell, 1978). Hayman and Parsons (1960) have presented limited data consistent with environmental and genomic stresses associated with recombination exceeding their individual effects, so that some combinations of genomic and environmental stress agents may generate variability permissive of rapid evolution in novel environments, but there is a need for additional critical experiments.

For quantitative traits, there is an increasing body of literature that has been reviewed elsewhere (Parsons, 1983, 1987). Following Schmalhausen's (1949) conclusion that variability is increased under extreme environments, Parsons (1987) concluded from evidence in marine organisms, *Drosophila*, the cruciferous plant, *Arabidopsis thaliana*, rodents and our own species that:

> "Both phenotypic and genotypic variability tend to be high under conditions of severe stress imposed by the physical and biological environments. Since stress increases the amount of genetic variation, it has the potential to maximize the rate and direction of evolutionary change. This applies directly to quantitative traits of importance in determining survival, and more indirectly at the level of genes controlling protein variation."

More recently, this conclusion has been extended from the ecological to the behavioral level (Parsons, 1988b). For example, in mice and rats, behavioral stress due to interactions between animals under crowding, leads to a number of traits showing increased additive genetic variability. These include pre-implantation mortality, litter size, relative adrenal weight and plasma corticosteroid levels indicating a direct change in endocrinological status in response to stress (Belyaev and Borodin, 1982).

Considering our own species, early hominid evolution produced a form adapted to the climate and biota of the hot savannah, but from *Homo erectus* onwards, hominid populations became adapted to an increasingly diversified combination of natural and cultural stresses leading to much of the biological and behavioral variation observed today. The model envisaged is a diversity of stresses promoting adaptation to new environments which create new stresses involving further adaptive responses (Baker, 1984). Important biological stresses include infectious diseases and altered food intake, such as ethanol ingestion which at stressful levels increases both behavioural and physiological variability (Martin *et al.*, 1985). Another example is maturity onset (non-insulin dependent) diabetes which can be regarded as a response to nutritional overload (Wirsing, 1985), and is one of the first diseases to appear following economic development towards that of affluent societies. Cultural stress leads to nutritional stress which is manifested as increased variability. As would be predicted, the major metabolic abnormalities of late-onset diabetes can be improved or completely normalized by a relatively short reversal of the urbanization process; a seven week period sufficed in a sample of diabetic Aborigines from north-western Australia (O' Dea, 1984).

THE COST OF STRESS

Even though stress, by increasing variability, can cause rapid evolutionary change in the most general sense, there is ultimately a cost. Artificial selection often reaches limits, not because genetic variability is exhausted, but because the fitness of the organisms so produced is lower than unselected organisms. For example, heavy metal resistance is easily increased by artificial selection in plants such as *Agrostis tenuis* (Bradshaw, 1984; MacNair, 1982). However, resistant plants tend to be competitively inferior to normal plants. A particularly interesting example comes from Mackay (1985) who studied a transposable element in *D. melanogaster* with the property of generating sufficient variation to enhance the response to selection for abdominal bristle score by 1.5 times that obtained when transposition was rare, but at the same time there is evidence for a fall in viability. This means that while rapid adaptation may occur in response to an environmental stress or selection for an extreme, ultimate limits to change will be determined by deteriorating fitness as the phenotype diverges from the usual environment of a population. In summary, stress is a time of change at many levels, but ultimately the exposure of genetic combinations not previously exposed to periods of selection by the environment means that a limit may be approached, at least in the short term.

Stress is an environmental probe for unmasking and studying variability. A general feature of stressed systems is an increase in energy expenditure, since following a severe perturbation, energy is diverted from maintenance and production to repair and recovery (Odum, Finn and Franz, 1979). One measure of the energy available to an organism at a given time is the adenylate energy change (AEC) expressed as the ratio:

$$\frac{[\mathrm{ATP}] + 1/2\,[\mathrm{ADP}]}{[\mathrm{ATP}] + [\mathrm{ADP}] + [\mathrm{AMP}]}$$

where [ATP], [ADP] and [AMP] represent amounts of adenosine triphosphate, diphosphate and monophosphate respectively (Atkinson, 1977; Hochachka and Somero, 1984). ATP basically supplies the power involved in the cost of living. Values of the AEC range from 0.9 when environmental conditions are optimal and non-stressed and active growth and reproduction occur, to 0.5 where the stress is so severe that viability losses are apparent even following a return to normal conditions (Ivanovici and Weibe, 1981). A wide range of environmental perturbations affect AEC's; these include nutritional stress, oxygen depletion, desiccation, heat and chemical stresses. This means that the AEC has a role as a metabolic indicator of the relative severity of an ecological stress, however, additional work is needed to assess how generally useful the AEC is, especially in marine organisms (Schafer and Hackney, 1987).

In any case, an organism should be able to increase its resistance to a range of environmental stresses via a reduction in metabolic rate. The evidence is scattered in the literature, and is reviewed in Hoffmann and Parsons (1989a, b). Some examples include (1) hibernation and diapause as adaptations to avoid environmental stress periods, (2) when metabolic rate is high at high temperatures, *D. melanogaster* shows reduced resistance to the stresses of anoxia, starvation, high ethanol concentration, and desiccation, (3) lines of *D. melanogaster* selected for postponed senescence were more resistant to environmental stresses and had lower metabolic rates than control flies (Service, 1987), (4) longevity is increased in *D. melanogaster* by reducing activity levels, (5) in the mole rat, *Spalax ehrenbergi*, the basal metabolic rate of races across the climatic zones of Israel decreases towards the desert where conditions are more stressful in terms of heat and desiccation (Nevo and Shkolnik, 1974), and (6) in the desert harvester ant, *Pogonomyrex rugosus*, the primary adaptation to xeric, or dry, conditions appears to be a lower-than-predicted metabolic rate affording a reduction in food requirements and respiratory water loss (Lighton and Bartholomew, 1988).

These considerations lead to two predictions:

1. reduced metabolic rate should be a feature of individuals genetically resistant to stress, and
2. positive genetic correlations are expected for resistance to different environmental stresses because genes decreasing metabolic rate are likely to increase resistance to a number of stresses.

Hoffmann and Parsons (1989a) took desiccation resistance as an appropriate stress trait to test these predictions. Substantial genetic variability had been previously demonstrated for desiccation resistance within and between populations from different geographic localities in *D. melanogaster* (Parsons, 1970, 1980). Populations from habitats where desiccation stress is sporadically severe tend to be more resistant than those from relatively benign habitats. Associations between desiccation resistance and habitat have also been found in other species, including mosquitoes (Machado-Allison and Craig, 1972) and various plants (Al-ani, Strain and Mooney, 1972; Derera, Marshall and Balaam, 1969).

Following the intensity of selection in the finch, *Geospiza fortis*, during a drought in the Galopagos Islands (Boag and Grant, 1981), an intense level of selection giving 85% mortality was chosen. Commencing with a large population collected in suburban Melbourne, Australia, the response to selection was rapid. Divergence of the selected lines from the controls occurred after four generations. Based upon LT 50's (the time taken for 50% of flies in a sample to succumb to stress) the mean of three selected lines after nine generations of selection (on females only) was 28.5 hours compared with a control mean of 18.6 hours. This difference of 9.9 hours is more than 50% of the resistance of the control lines. The average realized heritability over this period came to 0.64 which is extremely high, and is consistent with rapid responses to selection.

Such selected lines are ideal for investigations of correlated responses, especially with regard to the two above predictions. Hoffmann and Parsons (1989a, b) found that:

1. Stress-resistant strains have a reduced metabolic rate (as measured by O_2 uptake) and are less active behaviorally and with reduced fecundity, which are direct and predictable consequences of the fall in metabolic rate.

2. Stress-resistant strains are tolerant of starvation, toxic levels of ethanol and acetic acid, irradiation with exceedingly high doses of Co^{60}-gamma rays, and high temperatures.

The above predictions are therefore confirmed. An integrated approach to environmental stress and life-history variation therefore emerges by considering metabolic rate as the key trait, and using desiccation resistance as the environmental probe for magnifying genetic variability. Therefore, resistances to different stresses will often be correlated due to a common genetic mechanism relatable to metabolic rate. Genotypes with low rates of metabolism may then be favoured under a range of stressful conditions, since the cost of stress is reduced in them.

STRESSES TODAY AND IN THE FUTURE

In *The Genetical Theory of Natural Selection*, Fisher (1930) wrote:

> "If therefore an organism be really in any high degree adapted to the place it fills in its environment, this adaptation will be constantly menaced by any undirected agencies liable to cause changes to either party in the adaptation."

The scene today consists of increasing concentrations of CO_2 and less abundant atmospheric gases including chlorofluorocarbons (CFC's), all implying substantial exchanges between terrestrial systems and the atmosphere (Mooney *et al.*, 1987). Major depletions in stratospheric O_3 are now following (Cicerone, 1987). Collectively, these changes could increase world temperature by up to 5°C within the next century. Compared with prehistoric changes of similar magnitude this is exceedingly rapid, so that conditions of severe environmental stress are likely to affect the world's biota. Interestingly, while the current upsurge in interest in greenhouse gases is very recent, Lovelock (1979) writes of "the potential threat to the ozone layer arising from the nitrogen oxides and fluorocarbons". Here, I assume the rapid temperature increase and enquire about the likelihood of biological adaptation.

The dependence upon metabolic rate means that generalized stresses should be at least partially cumulative in effect. For example, global warming may be associated with simultaneous desiccation and pollution stress in some regions, and the potential and observed effects of multiple pollutants is becoming a matter of concern (Hinrichsen, 1986). In addition to the problem of obtaining large responses to selection for high temperature extremes (Parsons, 1989), such interactions would curtail the potential for adaptation to a simple temperature increase. However, an underlying dependence upon metabolic rate should permit predictions of populations and species most susceptible to stress. At the level of individual atmospheric pollutants, substantial additive genetic variability has been documented for SO_2 resistance in an increasing number of plant species, associated with a predictable tendency for the most resistant plants to be present in exposed habitats (Taylor and Murdy, 1975; Winner, Mooney and Goldstein, 1985). There are also reports of substantial genetic variability for levels of ozone damage (Pitelka, 1988). Since pollutants normally occur in combination, increasing reports of synergistic effects including acid rain (Hutchinson, 1984; Winner and Atkinson, 1986) indicate that adaptations should be governed by the genetic variability available not only for the effects of single pollutants, but also the effects of interacting combinations. Even partial genetic associations of pollutants with metabolic rate would exacerbate the effects of temperature change. This means that there is an urgent need for the experimental study of various stresses singly, and in combination, in insects such as *Drosophila* to provide a model of the potential effects of environmental stresses.

In considering the conservation of populations, the consequences of multiple stresses associated with global warming appear primary. Can the biota of the world adapt to such environmental shifts? Within species, this is a problem in quantitative genetics in understanding genetic response to stress. The potential for adaptive responses to a temperature increase is less than for more specific stresses which tend to have genetic architectures consistent with the potential for rapid change. Even so, changing the overall adaptive norm of a species is not easy, and in the case of temperature, reactions are strongly canalized to the range of existing populations (Schmalhausen, 1949). The difficulty is that all really new reactions of an organism are never adaptive. Geographically widespread species will be characterized by differing tolerances according to the habitat templet (Southwood, 1988), but ultimately the set of conditions under which metabolism is possible will be restrictive. Since a major consequence of stress adaptation is likely to be associated with a fall in metabolic rate, it follows that the scope of species to widen their adaptive norms in response to rapidly escalating stress is restricted, because of concomitant increasing metabolic costs (Hoffmann and Parsons, 1989a,b; Parsons, 1989).

The critical role of temperature relationships in determining the distribution limits and survival of organisms has been demonstrated. Biogeographers, population biologists, physiologists and biochemists representing many levels of biological organization have commented upon the pervasive and dominant importance of temperature in determining environmental relationships. More specifically, as temperature increases, the rate of chemical transformations accelerates. Very small temperature changes may have quite large effects on reaction rates. Furthermore, temperature changes frequently have substantial effects upon the equilibrium constants of biochemical reactions, especially those involving the reversible formation of noncovalent (or weak) chemical bonds. All of the diverse biological structures stabilized by weak chemical bonds share a common property of changing during the performance of their activities, i.e. , they are not rigid and invariant. A fine balance between lability and stability occurs. Hochachka and Somero (1984) write:

> "The reliance on weak chemical bonds dictates a sharp temperature dependence of macromolecular structure, and this dependence in turn makes the metabolic apparatus and the regulation thereof highly sensitive to temperature change. Likewise the rate

> effects of temperature will impact on the velocities with which metabolic flux occurs. For these reasons evolutionary changes directed toward coping with the effects of temperature are persuasive, as are the phenotypic changes observed in organisms confronted with short-term fluctuations in temperature."

Considerable metabolic complexity is therefore implied in temperature adaptations, which is consistent with rather restricted responses to selection for temperature extremes. This situation agrees with Mayr (1982b) who argues for evolutionary stasis largely due to an internal cohesion of the genotype perhaps sustained by as yet unexplored molecular mechanisms. Conversely it is not surprising that one of the main pleiotropic effects of mutants is an alteration, usually a reduction, in tolerance to a range of temperatures (Schmalhausen, 1949) simply because a mutant will disturb this internal cohesion.

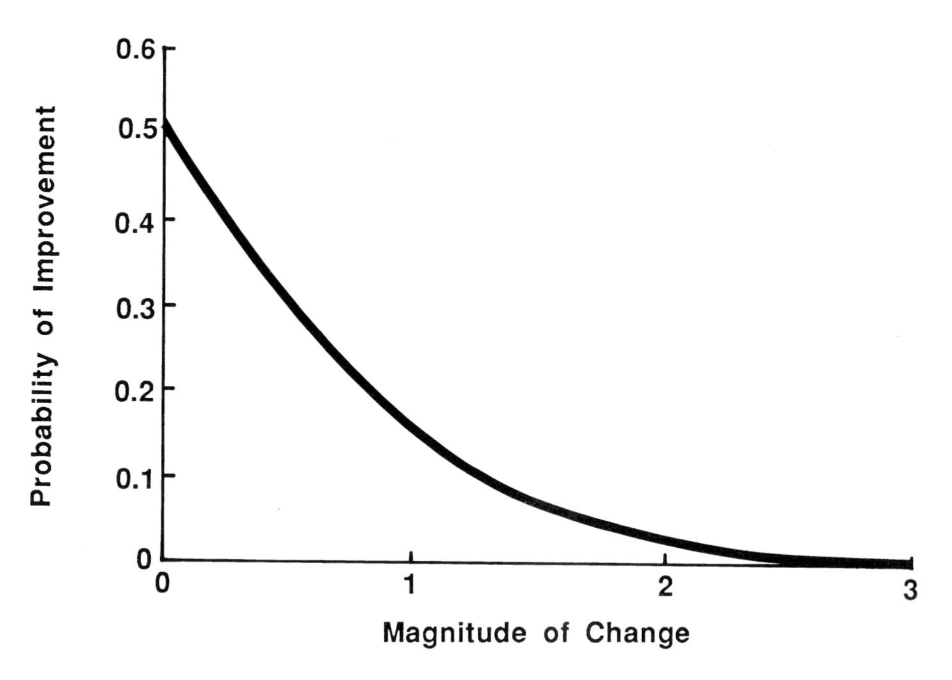

Figure 4. The relation between the magnitude of an undirected change and the probability of improving adaptation when the number of dimensions of the environment is large (after Fisher, 1930).

In recent years, genetically engineered organisms have been developed for a wide range of purposes. Most recombinant DNA work is laboratory based using organisms that are highly selected for adaptation and survival under restricted environmental conditions. New genes spliced into organisms tend to be characterized by destabilizing metabolic excesses and unforeseen phenotypic effects (Regal, 1988), which could include reduced resistance to temperature extremes. In other words the production of genetically engineered organisms may be metabolically costly.

Furthermore, there is likely to be an energetic load from the synthesis of extra macromolecules (Lenski and Nguyen, 1988). Such metabolic costs may not be a matter of initial concern because benign habitats are usually chosen for genetically engineered organisms. However, if these organisms were exposed to a changed environment then their conservation may be difficult. This is because the primary unaltered organism is the product of a series of past episodes of selection in variable environments where metabolic efficiency would be maximized.

Necessarily, there is the likelihood that the genetic background incorporating the novel function of the genetically-engineered organism will change in response to its environment. It would be difficult to prevent natural selection from acting upon genetically engineered organisms, whether this is considered desirable or not. It does follow, however, that new genetically-engineered organisms are unlikely to pose an environmental threat because they are already highly selected for survival under restricted conditions. Just as for novel mutations, the chance of a *de novo* genetic change being advantageous is low, under the variable environments to which organisms are normally exposed. The chances are greatest for very small changes (Fig. 4), but become increasingly small with increasing magnitude of change. Since the genetic engineer is usually dealing with large changes, the chances of immediate adaptation are low. It follows that a rapidly changing temperature regime as postulated for the immediate future would pose a greater threat to genetically-engineered organisms than to their progenitors which would be then preferentially conserved.

COMPETITION AND STRESS?

Two or more organisms coexisting in proximity may influence each other as they may share a resource that is in short supply. Such an influence is referred to as competition or interference, but here the term competition will be used. Competition implies a similarity in the requirements of the competing individuals. Thus Birch (1957) has written:

> "Competition occurs when a number of animals (of the same or of different species) utilize common resources the supply of which is short; or if the resources are not in short supply, competition occurs when the animals seeking that resource nevertheless harm one another in the process."

As the genetic resemblance between two individuals increases, so, too, will the similarity of their requirements, which in turn will be expected to lead to more intensive competition between those two individuals. It is reasonable, therefore, that the most intense and widespread competition would be expected to be within species and varieties, and that between-species competition would be of lesser importance. However, between-species competition would be expected to be most intense between those that are most closely related.

Especially in the evolutionary context, the relative importance of competition in natural populations continues to be debated extensively. It would be expected to develop maximally under circumstances of environmental stability rather than under fluctuating environments that tend to occur at the margins of species distributions. Under stable conditions, population size variation should be sufficiently dampened down such that population size would depend upon competitors, parasites and predators. These interactions all involve frequency-dependency (Clarke, 1979). The environmental stability envisaged implies a trend onwards K-selection of the ecologist, being a situation where Futuyma (1979) has argued for the importance of frequency-dependent selection and competition in determining community structure and genetic diversity. An equilibrium is therefore implied between populations and resources. However, at an extreme all populations are subject to ecological crises although the time scale may be lengthy in some situations (see Parsons, 1989 for discussion). At times of crises, equilibria and interactions would be disrupted, and in many cases destroyed. Hence under the stress paradigm of this article it can be argued that phenomena such as competition and frequency-dependency are second order. It is then not surprising that the debate concerning the importance of competition is continuing, extensive and occasionally heated.

One of the difficulties (Strong, 1983) is that much of orthodox competition theory is based upon observations in birds, lizards, and other vertebrates, which are only a small component of biotic diversity. Connell (1983) carried out an impressive analysis of published field experiments in an attempt to assess the relative importance of interspecific competition, covering 215 species and 527 experiments, and found competition in many of the organisms, in somewhat more than half of the species, and in about two fifths of the experiments. The prevalence of competition varied among organisms. Marine organisms showed consistently higher frequencies of competition than terrestrial ones as did large-sized organisms compared with smaller ones. Plants, herbivores, and carnivores showed similar frequencies of competition. An instability was indicated by the observation that the incidence of competition varied considerably from year to year and from place to place; in some cases this followed from stronger competition when resources were scarce. Considering different components of the biota, marine organisms may be exposed to less variability in physical characteristics than terrestrial ones, which is consistent with a higher incidence of demonstrable competition in marine species. Hence, a factor determining whether a species tends to show competitive relationships may be the degree to which it is subject to fluctuating stressful environments.

Stress resistance traits are generally highly heritable and adaptation to habitat characteristics occurs rapidly. Even though competitive ability can be increased in the laboratory by selection as shown by Mueller (1988) in *D. melanogaster*, it is a very low heritability trait. In contrast, heritabilities for environmental stress resistance tend to be at the upper end of the range. In terms of the potential for evolutionary change, stress differences among habitats should therefore have far greater influences than competitive relationships. Indeed, the instability of competition found by Connell (1983) could relate to stress differences altering the incidence of competition on both a temporal and a spatial basis.

Darwin (1859) greatly emphasized the significance of competition, its contribution to natural selection, and hence to evolutionary change:

> "As the species of the same genus usually have, though by no means invariably, much similarity in habits and constitution, and always in structure, the struggle will generally be more severe between them, if they come into competition with each other, than between species of distinct genera."

As Mayr (1963) notes, there immediately follows the problem that Darwin tended to refer to competition in much more dramatic terms than is usually assumed today, when even the definition of competition itself still remains a difficult problem. In some cases, Darwin's discussions of competition involve physical features of the environment.

> "Hence, as more individuals are produced than can possibly survive, there must in every case be a struggle for existence, either one individual with another of the same species, or with individuals of a different species, or with the physical conditions of life."

Eventually under conditions of extreme stress, he notes:

> "Not until we reach the extreme confines of life, in the Arctic regions or on the borders of an utter desert, will competition cease. The land may be extremely cold or dry, yet there will be competition between some few species, or between the individuals of the same species, for the warmest or dampest spots."

Darwin therefore appears to envisage a continuum where competition is exceedingly important in benign environments, becoming of minor importance under conditions of severe stress.

Notwithstanding Darwin's conclusions, competition appears to be of more transitory importance in evolutionary time than stress, which can be assumed to be universal. The low heritability of competitive ability compared with stress resistance provides the quantitative assessment behind this conclusion.

DISCUSSION

Schmalhausen (1949) wrote:

> "Basically, progressive evolution consists of a continuous acquisition of new reaction norms."

This is a succinct statement of the organism-environment interactions involved in evolutionary change at the phenotypic level. At the present time, the world is faced with selection of new reaction norms in response to the stress following global warming, which will increase the need for metabolic energy to counter these stresses. The potential rapidity of these changes transcends those characterizing the fossil record, where interpretations are difficult because assessments are retrospective.

Turner (1982) presented evidence for periods of rapid evolutionary change underlain by shifts in the frequencies of major genes and modifiers in a consideration of mimicry in Heliconius butterflies during a phase of race formation, which took place when the neotropical rainforests were fragmented and stressed during the cool, dry periods of the Pleistocene. Such periods of rapid change corresponding to periods of ecological change are described by Turner (1982) as "jerky" evolution.

Since, however, natural populations are continuously subjected to stress, especially at distributional margins, crossing the barrier to species extinctions would not appear to require a great escalation of stress. That is, a stress of greenhouse dimensions could lead to species extinctions as argued for the *Drosophila* fauna of the humid tropics of Australia (Parsons, 1989). The fossil record is characterized by mass extinctions which tend to occur simultaneously across many taxonomic and ecological groupings giving approximately the same extinction profiles. From their analysis of the fossil record, Raup and Boyajian (1988) conclude that extinction is physically rather than biologically driven.

Since the fossil record is mainly based upon morphological change, it is perhaps surprising that laboratory selection experiments on the effect of ecological stress upon morphological traits appear uncommon. However, this was an aim of the desiccation selection experiments discussed earlier. Three body measurements (head width, wing length, wing width) and wet weight were used to test for morphological changes associated with the selection response, however no significant effects occurred. At the anatomical level, all that can be said is to suggest that the rate of water loss via the spiracles was altered by selection, for example, a low metabolic rate may lead to flies keeping their spiracles closed longer because of a reduced demand for gaseous exchange (Hoffmann and Parsons, 1989a). To quote again from Schmalhausen (1949):

> "In other instances, physiological races are most pronounced and geographic races do not differ morphologically."

for which the above provides a laboratory model. Hence selection for desiccation resistance can lead to large changes, but the type of change appears to depend upon the organism. The consequences of the organism-environment interaction in responding to stress therefore depend ultimately upon the primary targets of selection in natural populations, and how these are affected by intense stress. It follows that periods of morphological stasis in the fossil record could be periods of substantial change at the physiological level. Only when a reaction norm crosses certain critical thresholds will observable morphological change occur, which is then likely to be rapid so giving "jerky" evolution observable at the morphological level, since a conalization zone (Waddington, 1956) may be disrupted.

Turning finally to our own species, it is not plausible to accept the view put forward by some that human evolution is stationary because of the "control" of earlier selective forces such as disease and malnutrition. In all organisms, an increasing array of substantial selective effects is being documented (Endler, 1986). It is more realistic to argue for differing selective forces at various times in human history, just as the change from the hunter-gatherer society to the initial development of urban settlements must necessarily have changed the selective forces operating upon populations. Because of increasing population density, the present important selective forces may well have more substantial behavioral components than in earlier times. If AIDS is to some extent a behavioral consequence of urban crowding, then we could be reverting to disease as a selective force. There is a major problem in the evolution of our own species, since the primary targets of selection are not fixed, but shift continuously in response to the ecological circumstance at any point of time. Behavioral stresses tend to be difficult to identify, but their study may become increasingly important as populations increase in size. The stressful scenario in which we find ourselves today means that the rate of human evolution could be accelerating, and a global climatic change may provide a powerful impetus.

SUMMARY

1. The interaction between organism and environment is central for understanding evolution in particular when environments are extreme, since experimental work, natural population studies, and distributional limits of species indicate that stress (in particular climatic and nutritional) is the norm in nature.

2. Phenotypic and genotypic variability tend to be high under severe physical and biological stress. This applies to mutation, recombination, and quantitative traits important in determining survival.

3. Stress is therefore an environmental probe for unmasking and studying variability, and this has been directly demonstrated for desiccation resistance in Drosophila, an important ecological phenotype that can be rapidly increased by directional selection. The cost in resistant flies is a reduction in metabolic energy, and predictably there are correlated responses for other stresses.

4. Since combinations of stress may be cumulative, and even synergistic in their effects, adaptation to the temperature increase proposed for the greenhouse affect may be complex. In any case, because climatic stress is common, major range expansions would be unlikely so that species extinctions would occur.

5. The release of genetically-engineered organisms should not normally pose problems because they would be unlikely to be as resistant to environmental extremes as primary unaltered organisms.

6. Because resistance to stress is highly heritable, it follows that competition has a lesser role than stress in evolution.

7. In interpreting the fossil record, major physiological shifts involving metabolic processes may occur in response to ecological stress without the necessity of simultaneous morphological change.

References

Alahiotis, S.N. 1982. Adaptation of Drosophila enzymes to temperature. IV. Natural selection at the alcohol-dehydrogenase locus. Genetica **59**: 81–87.

Al-ani, H.A. , Strain, B.R. and Mooney, H.A. 1972. The physiological ecology of diverse populations of the desert shrub *Simmondsia chinensis. J. Ecology* **60**: 41–57.

Andrewartha, H.G., and Birch, L.C. 1954. *The Distribution and Abundance of Animals.* Chicago: University of Chicago Press.

Atkinson, D.E. , 1977, *Cellular Energy Metabolism and its Regulation.*, New York: Academic Press.

Baker, P.T. 1984. The adaptive limits of human populations. *Man* (*N.S.*) **19**: 1–14.

Belyaev, D.K. and Borodin, P.M. 1982. The influence of stress on variation and its role in evolution. *Biol. Zentralbl.* **100**: 705–714.

Birch, L.C. 1957. The meanings of competition. *Am. Nat.* **91**: 5–18.

Boag, P.T. and Grant, P.R. 1981. Intense natural selection in a population of Darwin's finches (Geospizinae) in the Galapagos. *Science* **214**: 82–84.

Bouletreau, J. 1978. Ovarian activity and reproductive potential in a natural population of *Drosophila melanogaster. Oecologia* **33**: 319–342.

Boyer, J.S. 1982. Plant productivity and environment. *Science* **218**: 443–448.

Bradshaw, A.D. 1984. Adaptation of plants to soils containing toxic metals. *In Origins and Development of Adaptation*, pp. 4–19, CIBA Foundation Symp. 102. London: Pitman Books.

Busby, J.R. 1986a. Bioclimate prediction system Users Manual. Bureau of Flora and Fauna, Canberra.

Busby, J.R. 1986b. A biogeoclimatic analysis of *Nothofagus cunninghamii* (Hook) Oerst. in southeastern Australia. *Aust. J. Ecol.* **11**: 1–7.

Caughley, G., Short, J., Grigg, G.C., and Nix, H. 1987. Kangaroos and climate: An analysis of distribution. *J. Anim. Ecol.* **56**: 751–761.

Cicerone, R.J. 1987. Changes in stratospheric ozone. *Science* **237**: 35–42.

Clarke, B. 1979. The evolution of genetic diversity. *Proc. Roy. Soc. London Series B.* **205**: 453–474.

Clegg, M.T. and Allard, R.W. 1972. Patterns of genetic differentiation in the slender wild oat species *Avena barbata. Proc. Natl Acad. Sci. USA* **69**: 1820–1824.

Connell, J.H. 1983. On the prevalence and relative importance of interspecific competition: evidence from field experiments. *Am. Nat.* **122**: 661–696.

Darwin, C. 1859. *On the Origin of Species by Means of Natural Selection.* London: Murray.

Darwin, C., and Wallace, A. 1859. On the tendency of species to form varieties; and on the perpetuation of varieties and species by means of selection. *J. Linn. Soc. Lond.* (*Zool.*) **3**: 45–62.

Derera, N.F, Marshall, D.R. and Balaam, L.N. 1969. Genetic variability in root development in relation to drought tolerance in spring wheats. *Expl. Agric.* **5**: 327–337.

Endler, J.A. 1986. *Natural Selection in the Wild.* Princeton: Princeton University Press.

Fisher, R.A. 1930. *The Genetical Theory of Natural Selection.* Oxford: Clarendon Press.

Futuyma, D. 1979. *Evolutionary Biology.* Sunderland, Mass: Sinauer Associates.

Grell, R.F. 1978. A comparison of heat and interchromosomal effects on recombination and interference in *Drosophila melanogaster*. Genetics **89**: 65–77.

Hayman, D.L. and Parsons, P.A. 1960. The effect of temperature, age and an inversion on recombination values and interference in the X chromosome of *Drosophila melanogaster*. Genetica **32**: 74–88.

Hinrichsen, D. 1986. Multiple pollutants and forest decline. *Ambio* **15**: 258–265.

Hochachka, P.W. and Somero, G.N. 1984. *Biochemical Adaptation.* Princeton: Princeton University Press.

Hoffman, A.A. and Parsons, P.A. 1989a. An integrated approach to environmental stress tolerance and life—history variation: Desiccation tolerance in *Drosophila. Biol. J. Linn. Soc.* **37**: 117–136

Hoffman, A.A. and Parsons, P.A. 1989b. Selection for increased desiccation resistance in *Drosophila melanogaster*: Additive genetic control and correlated responses to other stresses. *Genetics* **122**: 837–845

Hutchinson, T.C. 1984. Adaptation of plants to atmospheric pollutants. In *Origins and Development of Adaptation*, pp. 52–72, CIBA Foundation Symp. 102. London. Pitman Books.

Ivanovici, A.M. and Wiebe, W.J. 1981. Towards a working 'definition' of stress: a review and critique. In *Stress Effects on Natural Ecosystems*, ed. G.W. Barrett and R. Rosenberg, pp. 13–27. New York: John Wiley.

Kohlmann, B., Nix, H. and Shaw, D.D. 1988. Environmental predictions and distributional limits of chromosomal taxa in the Australian grasshopper *Caledia captiva* (F). *Oecologia* **75**: 483–493.

Lenski, R.E. and Nguyen, T.T. 1988. Stability of recombinant DNA and its effect on fitness. *Trends in Ecol. and Evol.* **3**: 518–520.

Lighton, J.R.B. and Bartholomew, G. A. 1988. Standard energy metabolism of a desert harvester ant, *Pogonomyrex rugosus:* Effects of temperature, body mass, group size, and humidity. *Proc. Natl. Acad. Sci. USA* **85** : 4765–4769.

Lovelock, J.E. 1979. *Gaia : A New Look at Life on Earth.* Oxford : Oxford University Press.

Lucchesi, J.C. and Suzuki, D.T. 1968. The interchromosomal control of recombination. *Ann. Rev. Genet.* **2**: 53–86.

Machado-Allison, C.E. and Craig, G.B. 1972. Geographic variation in resistance to desiccation in *Aedes aegypti and A. atropalpus* (Diptera : Culicidae). *Ann. Ent. Soc. America* **65**: 542–547.

Mackay, T.F.C. 1985. Transposable element–induced response to artificial selection in *Drosphila melanogaster. Genetics* **111**: 351–374.

MacNair, M.R. 1982. Tolerance of higher plants to toxic materials. *In Genetic Consequences of Man-Made Change*, ed J.A. Bishop and L.M. Cook, pp. 177–207. London: Academic Press.

Martin, N.G., Oakeshott, J.G., Gibson, J.B., Starmer, G.A., Perl, J. and Wilkes, A.V. 1985. A twin study of psychomotor and physiological responses to an acute dose of alcohol. *Behav. Genet.* **15**: 305–347.

Mayr, E. 1963. *Animal Species and Evolution.* Harvard: Belknap Press.

Mayr, E. 1982a. *The Growth of Biological Thought: Diversity, Evolution and Inheritance.* Harvard: Belknap Press.

Mayr, E. 1982b. Speciation and macroevolution. *Evolution* **36**: 1119–1132.

McClintock, B. 1951. Chromosome organization and genic expression. *Cold Spring Harbor Symp. Quant. Biol.* **16**: 13–47

McClintock, B. 1984. The significance of responses of the genome to challenge. *Science* **226**: 792–801.

Mooney, H.A., Vitousek, P.M. and Matson, P.A. 1987. Exchange of materials between terrestrial ecosystems and atmosphere. *Science* **238**: 926–932.

Mueller, L.D. 1988. Evolution of competitive ability in *Drosophila* by density-dependent natural selection. *Proc. Natl. Acad. Sci. USA* **85**: 4383–4386.

Neel, J.V. 1941. A relation between larval nutrition and the frequency of crossing over in the third chromosome of *Drosophila melanogaster. Genetics* **26**: 506–516.

Nevo, E., Beiles, A. and Ben-Shlomo, R. 1984. The evolutionary significance of genetic diversity: Ecological, demographic and life history correlates. In *Evolutionary Dynamics of Genetic Diversity*, ed G.S. Mani, pp. 13–213. Berlin: Springer–Verlag.

Nevo, E., and Shkolnik, A. 1974. Adaptive metabolic variation of chromosome forms in mole rats Spalax. *Experienta* **30**: 724–726.

Nix, H.A. 1981. The environment of Terra Australis. In *Ecological Biogeography of Austsralia*, ed. A. Keast, pp. 103–133. The Hague: Junk.

O'Dea, K. 1984. Marked improvement in carbohydrate and lipid metabolism in diabetic Australian Aborigines after temporary reversion to traditional lifestyle. *Diabetes* **33**: 598–603.

Odum, E.P., Finn, J.T. and Franz, E. 1979. Perturbation theory and the subsidy-stress gradient. *Bio Science* **29**: 349–352.

Osmond, C.B., Austin, M.P., Berry, J.A., Billings, W.D., Boyer, J.S., Dacey, J.W.H., Nobel, P.S., Smith, S.D. and Winner, W.E. 1987. Stress physiology and the distribution of plants. *BioScience* **37**: 38–48.

Parsons, P.A. 1970. Genetic heterogenity in natural populations of *Drosophila melanogaster* for ability to withstand desiccation. *Theoret. Appl. Genet.* **40**: 261–266.

Parsons, P.A. 1980. Isofemale strains and evolutionary strategies in natural populations. *Evol. Biol.* **13**: 175–217.

Parsons, P.A. 1981. The evolutionary ecology of Australian Drosophila: a species analysis. *Evol. Biol.* **14**: 297–350.

Parsons, P.A. 1987. Evolutonary rates under environmental stress. *Evol. Biol.* **21**: 311–347.

Parsons, P.A. 1988a. Evolutionary rates: effects of stress upon recombination. *Biol. J. Linn. Soc.* **35**: 49–68.

Parsons, P.A. 1988b. Behavior, stress and variability. *Behav. Genet.* **18**: 293–308.

Parsons, P.A. 1989. Environmental stresses and conservation of natural populations. *Ann. Rev. Ecol. Syst.* **20**: 29–49.

Pitelka, L.F. 1988. Evolutionary responses of plants to anthropogenic pollutants. *Trends in Ecol. Evol.* **3**: 233–236.

Plough, H.H. 1917. The effect of temperature on crossing over in *Drosophila. J. Exp. Zool.* **24** : 148–209.

Raup, D.M. and Boyajian, G.E. 1988. Patterns of generic extinction in the fossil record. *Paleobiology* **14**: 109–125.

Regal, P.J., The adaptive potential of genetically engineered organisms in nature. *Trends in Ecol. Evol.* **3**: 536–538.

Rosewell, J. and Shorrocks, B. 1987. The implication of survival rates in natural populations of *Drosophila*: capture–recapture experiments on domestic species. *Biol. J. Linn. Soc.* **32**: 373–384.

Schafer, T.H. and Hackney, C.T. 1987. Variation in adenylate energy charge and phosphoadenylate pool size in estuarine organisms after an oil spill. *Bull. Environ. Contam. Toxicol.* **38**: 753–761.
Schmalhausen, I.I. 1949. Factors of Evolution. Philadelphia: Blakiston Company.
Schultz, J. and Redfield, H. 1951. Interchromosomal effects on crossing over in *Drosophila. Cold Spring Harbor Symp. Quant Biol.* **16** : 175–197.
Service, P.M. 1987. Physiological mechanisms of increased stress resistance in *Drosophila melanogaster* selected for postponed senescence. Physiol. Zool. **60**: 321–326.
Southwood, T.R.E. 1988. Tactics, strategies and templates, *Oikos* **52** : 3–18.
Stanley, S.M. and Parsons, P.A. 1981. The response of the cosmopolitan species *Drosphila melanogaster* to ecological gradients. Proc. Ecol. Soc. Aust. **11**: 121–130.
Strong, D.R. 1983. Natural variability and the manifold mechanisms of ecological communities. *Am. Nat.* **122**: 636–660.
Taylor, G.E. and Murdy, W.H. 1975. Population differentiation of an annual plant species, *Geranium carolinianum* L, in response to sulphur dioxide. Bot. Gaz. **136**: 212–215.
Turner, J.R. G. 1982. Darwin's coffin and Doctor Pangloss - Do adaptationist models explain mimicry? In *Evolutionary Ecology*, ed B. Shorrocks, pp. 313–361 Oxford: Blackwell.
Waddington, C.H. 1956. *Principles of Embryology*. London: George Allen and Unwin.
Wake, D.B. 1986. Foreward to *Factors of Evolution* by I.I. Schmalhausen. Chicago: University of Chicago Press.
Wallace, A.R. 1876. The *Geographical Distribution of Animals*. London: Macmillan.
Watson, J.D. and Crick, F.H.C. 1953. The structure of DNA. *Cold Spring Harbor Symp. Quant. Biol.* **18**: 123–131.
Winner, W.E. and Atkinson, C.J. 1986. Absorption of air pollution by plants and consequences for growth. *Trends in Ecol. Evol.* **1**: 15–18.
Winner, W.E., Mooney, H.A. and Goldstein, R.A. (eds) 1985. *Sulphur Dioxide and Vegetation: Physiology, Ecology and Policy Issues*. Stanford University press.
Wirsing, R.L. 1985. The health of traditional societies and the effects of acculturation. *Curr. Anthrop.* **26**: 303–322.

4. Mathematics and Biology

Alan L. Mackay

Professor Emeritus, Birkbeck College, University of London, U.K.

Mathematics is a way of stating precise relationships which may represent models of the real world. It uses numbers, symbols and pictures. Computation has enabled us to see the consequences of these relationships in, for example, ecology, morphology and genetics, where simple local connections can produce complex global patterns. The idea of neural networks, which model certain aspects of the way the brain works, has led to new types of computing which may be more suitable for biological problems. Most recently, computer graphics have greatly increased the intelligability of mathematical arguments and results. Biology has been slow to incorporate mathematical analysis, but the attempt by D'Arcy Thompson in his "Growth and Form" can now be taken forward. The basic difficulties are not with the mathematics, but with the ways in which biological problems are projected into mathematical form and, with how, after processing, the answers are interpreted back into the real world, that is, with 'reductionism'.

INTRODUCTION

Mathematics is a mode of thought, perhaps the very rules of thought itself, which George Boole believed he had discovered, or the language of thought as exact as we can make it. Mathematics can be applied to any field of study, physics being the most successful area. As biology becomes better understood, so aspects of it can be discussed more exactly, but biology is extremely complex compared with physics. There is no special area "mathematics and biology" but new parts of mathematics can be developed to handle specific problems of biology. *"The unreasonable effectiveness of mathematics"* in representing nature has been commented on. However, this effectiveness has been greatest in physics. Mathematics has been developed first for physics and not for biology. It is to be expected that the great current success of biological science, principally because of the rise of molecular biology, will stimulate new mathematics for biological problems at levels above that of molecules.

To some extent this has happened, first, for example, in the area of catastrophe theory and more recently in the topic of *cellular automata.* Hitherto mathematics has concentrated on linear effects, where response is directly proportional to stimulus. Biology involves essentially non-linear effects and rules. Handling these is largely a matter of computational power, which has only recently become widely available.

Is mathematics invented or discovered? We can now see an answer to this question through the example of cellular automata. Starting from a set of rules, a system (of the fourth of S. Wolfram's classes [Wolfram *et al.*, 1984]) passes from one state to another without repetition ... if the state were exactly repeated then it would continue in the same way as before and would thus be cyclic. Mathematics as a set of propositions starts, as in Euclid's "Elements", with a number of axioms and explores all the results of their possible combinations. In this sense mathematics is discovered, but the axioms have to be invented, and the main difficulty is in finding all their consequences. In the fourth class of cellular automata there is no quicker way of finding the consequences of a set of rules than simply to track the behaviour of the system itself. There is no better model or algorithm. In either case, there is plenty of room for creativity.

The computer has recently played a vital part in exploring the consequences of simple axioms and has shown the unimagined complexity which can arise, for example in the Mandelbrot set, from a few mathematical rules.

ATOMICITY AND HIERARCHY

It is the basic principle of biology that everything is made of atoms. Richard Feynman in his famous Lectures on Physics wrote: "Everything is made of atoms. That is the key hypothesis in all of biology. The most important hypothesis in all of biology, for example, is that everything that animals do, atoms do. In other words, there is nothing that living things do that cannot be understood from the point of view that they are made of atoms acting according to the laws of physics. This was not known from the beginning, it took some experimenting and theorizing to suggest this hypothesis, but now it is accepted, and it is the most useful theory for the production of new ideas in the field of biology."

New experimental techniques, particularly the scanning tunnelling microscope and high resolution electron microscopy generally, enable us to see and manipulate individual atoms. The hypothesis of atoms is now extremely secure. Molecular biology stands on the understanding of the properties of arrangements of atoms.

Below the level of atoms there are subatomic particles, such as the nucleons and quarks. However, just as we discount the influence of the stars on human behavior (and vice versa), so the subatomic world is decoupled from the scale of biology. The ranges of energy, time and size are very different. We see matter as a hierarchy with structures at different levels. These levels are only very slighly coupled to each other. The interactions are small but not in every case negligable. The key practical problem is the degree to which weakly coupled systems can be considered separately. This estimation of separability is a key matter for everyday life as well as for science. We may see distinct levels at atom; cell; organism; society.

Fractal mathematical patterns, for example the Mandelbrot set, although they may resemble biological hierarchies, are not necessarily good models for biological systems. This is because the patterns have infinitely detailed structures and subdivision produces more and more detail, whereas in biological system, subdivision stops at the level of atoms.

REDUCTIONISM

The central problem of mathematics and biology is that of reductionism. Reductionism is frequently used simply as a term of abuse. Science is said to be reducing the complexity of human beings to assemblies of atoms, neglecting human characteristics, and so on. It is often contrasted with 'holism' which is supposed to treat organisms and life itself as a whole.

Mathematics, language and pictures are the three main tools which we have for handling the data of experience and for communicating it to others. All these can be regarded as models, simplifications of the real situation with which we can work to think out what we should do next. Every creature has a model of its environment and itself with which it operates. This model cannot be a complete model but must be a simplification where essential features are included and many other things are omitted.

Behind an inn, in the town of Burton-on-the-Water in England, there is a model of the town, on perhaps one hundredth scale. In the model one can see the inn and behind it the model on a still smaller scale, which can be seen to contain the inn and so on in what should be an infinite regress. The models which organisms use cannot be like that because sooner or later we come down to the atomic level and cannot represent more detail. It used to be thought that the spermatozoon contained a homunculus, complete in all details, but smaller. Now we know that the fertilised cell contains instructions for its development and not a scaled down model. Moreover the instructions are written in the same atoms which form the structure they are describing.

For any kind of forecast or communication a real situation has to be projected into the model space. In the model space hypothetical operations are carried out, scenarios are tested and the behaviour of the model is predicted. These results are then restored to the real space and the consequences for the real situation are assessed. We either act on the prediction of the thought experiment if it seems reasonable or if not, we correct the model by re-assessing the significance of various features, including neglected features found to be important and omitting others which appear negligable. We may assess the reliability of the model by running it backwards or sideways to see if it gives predictions corresponding to what is already known.

The operation of projection is a very general one. In every projection, dimensionality is lost. Only selected information can be preserved. The essence of a good model is that the information which is significant for a particular application is preserved and everything else is not taken into account. Thus, the whole real situation cannot be restored or resurrected from the model (The purest case of projection and restoration can be seen in the generalised inverse of a matrix).

When we describe a situation in a natural language we project it into the language space. We manipulate it in this space; that is, discuss and analyse it, estimating what may happen. Then we translate our conclusions back to real life. We can only apply our conclusions from their projection in language by the addition of a great deal of further information which we may already have about real life.

The target of holism is quite unrealistic. The only full model of a complex organism is itself. The organism too cannot be separated from its environment and this cannot all be modelled. There is just no room.There is no alternative to choosing a part of the universe and, for the purposes of a particular analysis, neglecting its connections to other parts. Although we may see the universe structured in almost separable levels, the interactions between levels will, under particular conditions, become critical and non-negligable. Micro-events in the level below may critically turn into macroscopic events. Leo Tolstoi, in his epic War and Peace, discusses this in connection with the role of the individual in history. Large-scale events may change the 'climate' in which small-scale events occur. Thus, systems, including the universe and life itself, have to be studied theoretically and experimentally on all levels simultaneously, so that we can come to reasonable estimates as to what degree sub-systems can be seen as coupled.

MATHEMATICS FOR BIOLOGY

It is not necessary or possible to give an account of all the ways in which mathematics has been applied in biology. As biology has become a science so it has adopted scientific characteristics and the chief of these is the necessity for exact expression and mathematics is the means for such expression.

Euclidean mathematics was totally unsuitable for the biology of the time. The Greeks were preoccupied with whether the square root of 2 could be represented as the ratio of two integers. Plato is supposed to have said: "He is unworthy of the name of man who is ignorant of the fact that the diagonal of a square is incommensurable with its side" (Constance Reid, *A long way from Euclid*, p13). The closed logical Euclidean system was far from the squashy observations of natural history.

A few stages might be mentioned; William Petty (1623-1687) first began to apply statistics to human affairs in examining human mortality statistics. This became the foundation of the life insurance business. A.J. Lotka, (in that business) who must be reckoned as one of the founders of the application of mathematics to biology examined population statistics and demonstrated also the hazards of prediction. He wrote [Lotka, 1956]:

> "It would be a strange trick of fate if we, the most advanced product of organic evolution, should be the first of all living species to forsee its own doom."

However, the doom which Lotka foresaw in 1945 was that of human underpopulation, which now appears unlikely.

Euclidean geometry was, however, good for the description of static form although it was a long time before D'Arcy Thompson produced his classic book on 'Growth and Form'. He said, for example:

> "Cell and tissue, shell and bone, leaf and flower, are so many portions of matter, and it is in obedience to the laws of physics that their particles have been moved, moulded and conformed.They are no exception to the rule that God always geometrizes.Their problems of form are in the first instance mathematical problems, their problems of growth are essentially physical problems and the morphologist is, *ipso facto,* a student of physical science." [Thompson, 1917]

Lotka applied essentially linear mathematics to dynamic systems and the great advances of recent times have essentially been the analysis of the properties of non-linear systems. Mathematics was constrained to deal only with problems which could be solved with the methods available and non-linear systems were simply too difficult for pen and paper methods. It is only when computing power became available that the complexity of life could be even approached. Now concepts of 'complexity', 'chaos', cellular automata, informational systems, neural networks, preoccupy both biologists and computer scientists. In some ways mathematics has become too important to be left to the mathematicians. In order to understand the meaning of mathematical relationships it has become necessary to have developed computer graphics as an interface between formulae and the human mind. Now, even mathematicians, who often scorned the computer, have in many cases been captivated by its possibilities.

THE COMPUTER AND ITS OFFSPRING

The computer is a powerful metaphor for a living system but is not as close as it will become when highly parallel, self-organising computers appear [Moravec, 1988]. It must be taken as obvious that the computer now plays a decisive role in all fields. It is not only a tool for doing mathematics and is an amplifier of human capabilities both intellectual and physical, but it is a generalised model which can be employed in modelling almost anything.

INFORMATIONAL SYSTEMS

There are several stages in dealing with biology, first describing the static forms and then finding the laws of growth and function. The whole system evolves and we may discern many levels in the interaction of matter and information, culminating in the human cultural system. After this we may consider the origin of life as a stage in the evolution of matter and the possibilities of creating life in the present form or with different ingredients (such as microelectronics).

With the discovery of the genetic code and the immense complexity of the connections between DNA, proteins and the materials synthesised by proteins, we now see that the concepts of hierarchic structure discussed above are not sufficient. Structure, in the sense of the arrangement of atoms, is not just structure, as it might be when considering geology. Some structures are not just structures but describe other structures in another language. The situation is as if a visitor from outer space had entered the British Museum, examined the halls full of artifacts, and had then passed into the Library. What is the significance of the minute patterns of ink on the pages of the books? Certain arrangements of atoms are not just structures but describe and control the arrangements of other atoms. The connections between various levels of the hierarchy may be very complex and more important than they might at first signs appear to be.

All these stages are the subject of mathematisation, not as an end in itself, but by way of describing observed relationships most concisely and in working out their consequences.

Ideas of cellular automata have become essential for understanding biological and other systems. A living system embodies its own description in another language. The joint system evolves by the dialectical interaction of the object and its description. Each organism has a model, however vestigial, of itself and of its environment. The evolution of a system is its history.

GOD(S) AND BIOLOGY

What then is biology? Biology is the science of life and of living material. And what is life? Is a flame life? In a sense yes, although life is now associated with the element carbon, but that is perhaps not necessary. What distinguishes the living from the dead? We will not speculate here on what other systems with life-like behavior may be conceivable. We know only one kind of life and all life that we know belongs to the same unitary system based on the atoms carbon, hydrogen, oxygen, nitrogen, phosphorus, sulphur and a larger number of trace elements.

The form of life on Earth arises at the atomic level. The bond joining the atoms of life have energies of a few electron volts. That is, the structures and processes at levels below the outer shells of the atom play no significant part in the operation of life (it is for this reason that radioactive tracers are so important in following biological processes).

Life uses, not only the ready-made components of biology, but any other processes of nature which can be incorporated. It has to live with all the properties of matter, like gravity and surface tension as well as chemistry.

We are still working out the consequences of atomicity. The properties of atoms have given rise to an informational systems genetics, which has made life possible. Living systems have developed their own models of themselves and their environments. These models are made of atoms which thus represent the states of other atoms. The operation of these models is consciousness. At a still higher level, organisms conscious of this consciousness, have represented their models by writing, language and other external systems, but again all made of atoms. Organisms are now well on the way to creating, out of atoms, auxiliary brains in the shape of computers. It is quite impossible to have predicted that atoms would have all these complex emergent properties.

How we analyse biology depends on our overall view of the universe, ourselves and the creatures in it. There are two main views, the scientific and the religious. The religious views occur in several mutually irreconcilable and mostly local versions, whereas there is only one corpus of scientific knowledge with its users and contributors somewhat tenuously distributed over the world. The clash between these views is becoming sharper with the increased use of both religion and science as devices for controlling, as well as for understanding, human society.

In many societies the religious view has triumphed politically and it has become impossible to conduct science in them as an integral part of intellectual life. Children are imprinted with local religious views at an age which makes later scientific education difficult. In India, the materialist sects, the Lokayata or Charvaka, were suppressed at an early stage [Chattopadhyaya, 1959], although now India is a secular state with no offical religion. Even in present-day America large groups believe in the supernatural origins of life and science is under constant pressure from irrationalism. In Britain, even Mrs.Thatcher, when Prime Minister, despite (or perhaps because of) having received a scientific education at Oxford University, could proclaim that "We were made in God's own image" (Guardian, 23 May, 1988, p.88). An increasing number of nation states have become theocratic with officially inculcated religious views. A most interesting examination of the interaction of physics and religion, in India has been given by Lee Siegel [1991] who discusses the physical basis of miracles and the psycho-physics of belief.

It is impossible to understand the development of biology (or of politics for that matter) without considering the background of the continuing conflict between theistic and atheistic views of life as to which embodies the truth. But "what is truth?" and like Pilate we cannot wait here for an answer nor consider the political consequences if the religious views of theocratic states are found not to be true. Scientific statements, on the other hand, are always ready for reassessment by any tests for truth and if they fail, they are removed from the corpus of science.

It is indeed a curious paradox that both views of life have emerged from the properties of atoms. Evolutionary pressures will eventually decide between them and it is not even evident that it will be the scientific version which will prove to have the greater survival value.

References

S. Wolfram, J.D. Farmer and T. Toffoli [1984]. Cellular Automata, Physica D, Nos. 1 and 2.

A.J. Lotka. *Elements of Mathematical Biology* [1956] originally Elements of Physical Biology, 1924 Dover.

D'Arcy W. Thompson [1917]. *On Growth and Form.* Cambridge University Press.

H. Moravec [1988]. *Mind Children: The Future of Robot and Human Intelligence.* Harvard University Press.

Debiprasad Chattopadhyaya [1959] *Lokayata – A study in Ancient Indian Materialism.* People's Publishing House, New Delhi.

L. Siegel [1991]. *Nets of Magic.* University of Chicago Press.

Suggested Reading

R. Beale and T. Jackson [1990]. *Neural Computing,* Institute of Physics, Bristol.

R. Dawkins, *The Blind Watchmaker* Longman, London, 1986. *The Selfish Gene.* Oxford Univ. Press, 1976.

Ernst Haeckel [1866] *Generelle Morphologie der Organismen,* Berlin. Georg Reimer.

S. Hildebrandt and A.J. Tromba [1985] *Mathematics and Optimal Form.* San Francisco: Freeman.
Robert M. May [1973] *Stability and Complexity in Model Ecosystems.* Princeton University Press.
J.R. Newman [1956] *The World of Mathematics.* 4 vols., New York. Simon and Schuster.
J. Maynard Smith [1972]. *On Evolution.* Edinburgh Univ. Press.
C.H. Waddington [1969] *Towards a Theoretical Biology.* 4 vols. Edinburgh University Press.
E.O. Wilson [1975] *Sociobiology: the new synthesis.* Harvard Univ. Press.
A. Winfree. *The Geometry of Biological Time.* San Francisco: Freeman.

5. The Development of Order in Biology

Gordon A. Rodley

Centre for Peace and Conflict Studies, The University of Sydney, Mackie Building K01, NSW 2006 Australia.

INTRODUCTION

The amazing levels of complexity and organisation of living systems present interesting challenges to scientists. How could such detail as the optical precision of the eye or the fine tuning of the molecular components of the photosynthetic system arise from chemical processes? On the one hand, the development of such systems may appear to defy the general laws of thermodynamics governing chemical energetics. On the other, it is now recognised that certain chemical systems can readily and spontaneously become ordered. Yet the details of the processes involved remain tantalisingly obscure. The primary reason for this is the non-equilibrium and non-linear character of them. From mathematical and other viewpoints these are particularly difficult to analyse. Nonetheless, with the development of concepts such as 'dissipative' structures to identify conditions for ordering, and the availability of high-speed computers to test complex models, considerable progress has been made. New and striking results have been obtained concerning pattern formation in complex physical and chemical systems. Indeed, as Mandelbrot has remarked; "It has been a surprise that complex structures and features arise from systems characterised by very simple rules". Interesting examples surround us in everyday life, from cloud formations to frost patterns on windows. 'Order-out-of-chaos' has gained conversational currency. Systems are referred to as 'self-organising' and ordered features as 'emergent'. None-the-less, we remain puzzled by ordering.

THE NATURE OF ORDER

Order is essentially arrangement. It involves the way matter entities fit together to form patterns. These may be either spatial or temporal (Fig. 1) or a combination of both, as in living systems. Spatial ordering is recognised by the regularity or symmetry of the fitting together of component entities. Temporal ordering involves time-wise linkage of reactions to produce effects like oscillatory behavior. While its origin is obscure, ordering is ubiquitous. This is particularly the case for systems involving a delicate balance between order and disorder, since these produce the greatest richness of ordering possibilities (Fig. 2). Such a situation probably constitutes the single most important principle governing the development of life. It identifies those conditions yielding the broadest connectivity between a system and its environment. Moreover, it provides an arrow of time along which evolution may progress. The essential requirements are an underlying element of symmetry formation in the system, coupled with a capacity for randomness or disorder to occur.

For the solid state this is seen in crystallisation processes such as dendritic growth. Molecular entities pack in an ordered way within basic repeat units (unit cells), with the latter tending to link disorderly. Under appropriate growth-rate conditions this combination leads to macroscopically ordered growth patterns, as displayed by snowflakes. Variation of shape for the latter exemplifies just how diverse such patterns can be, within a basic symmetry class. Even more disordered arrays, known as fractal patterns, also display macroscopic ordering. The remarkable feature in these cases is the existence of the same basic shape at different levels of magnification. Frost patterns are an example of fractal patterning. In solution chemistry in-between order arises when fixed ('ordered') reaction pathways combine with appropriate degrees of diffusional 'disorder' to produce non-equilibrium steady states. Oscillatory behavior, where a system becomes kinetically locked into a cyclical set of reactions, is the more striking feature of order displayed by dynamic systems. This arises when reaction pathways are available to switch the system from one steady state to another.

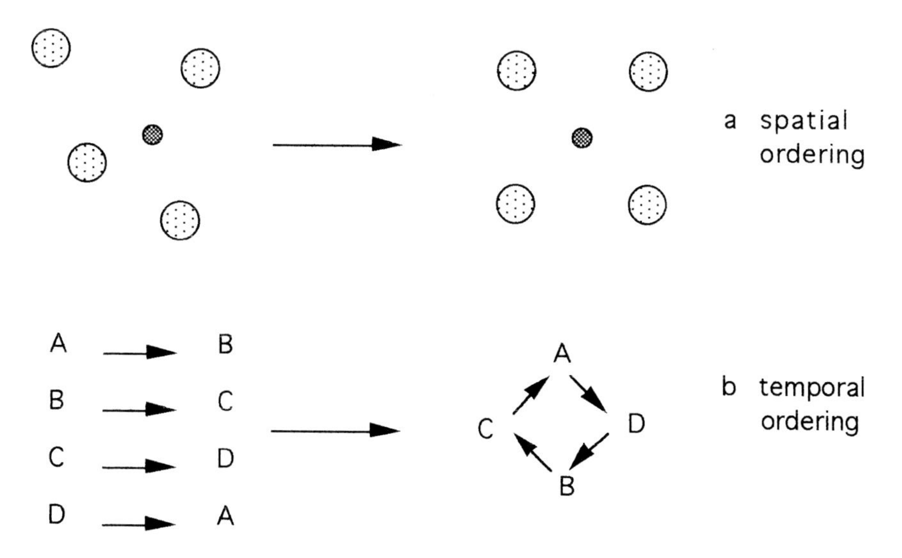

Figure 1. Ordering processes

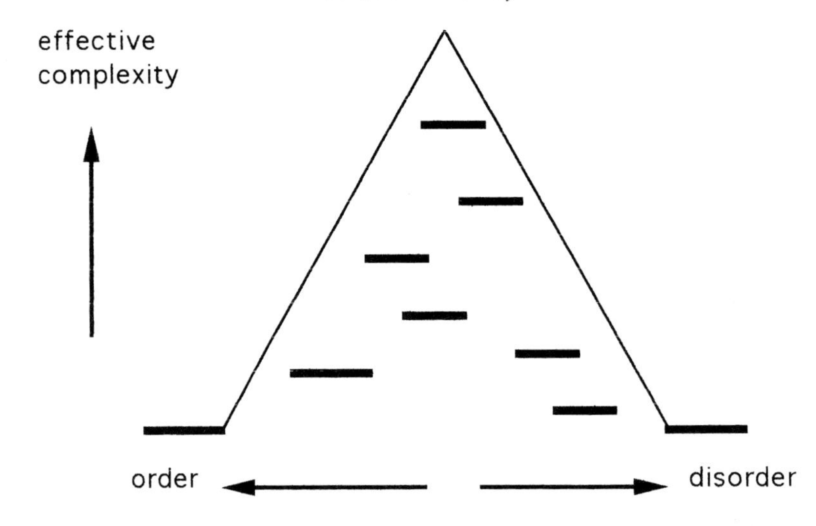

The development of order may also be understood in a general way in terms of chemical or matter flow being inhibited in some manner. Patterns may emerge for an appropriate balancing of flow and inhibition. The general picture is that whenever energy dissipation is hindered ordering of some kind will result. Thus, in heating a liquid between two horizontal plates, a point is reached where normal conduction cannot cope with heat dissipation. At this stage the system is forced into an ordered pattern of coherent convection motion, known as Benard structure (Fig. 3a). Water flowing down a river, but impeded in its flow by a rock, forms wave patterns around the rock (Fig. 3b). In both of the latter cases the forces of flow induce molecular units to stick together to form discrete packages. The same imagery may be applied to chemically reacting systems. Beyond these kinds of ordering is yet a more interesting possibility; the combination of ordered systems of different types, undoubtedly of considerable significance in the emergence of life.

With new understanding of this kind, there is a greater preparedness to consider the order of living systems to have arisen 'naturally'. Our tendency to think otherwise may be more a reflection of ignorance about the potentialities of matter than of an actual situation. This issue has been succinctly analysed in 'The Blind Watchmaker' by Richard Dawkins. The argument often raised (referred to by Dawkins as 'The Argument from Personal Incredulity') is that the time available for the remarkable development of an entity like the eye has been too short. What Dawkins and, more recently, others have presented, are computer simulations to illustrate the apparent ease with which ordered structures may evolve. This is part of a broader issue about our perceptions of ordering. The approach taken here is to further identify 'natural' features of matter underlying the production of spatial and temporal order.

However, an important emphasis to make is that ordering at the biological level is a function of the total matter world. It involves atom and molecular formation at a fundamental level, thermodynamic considerations, inter-stellar gas cloud and

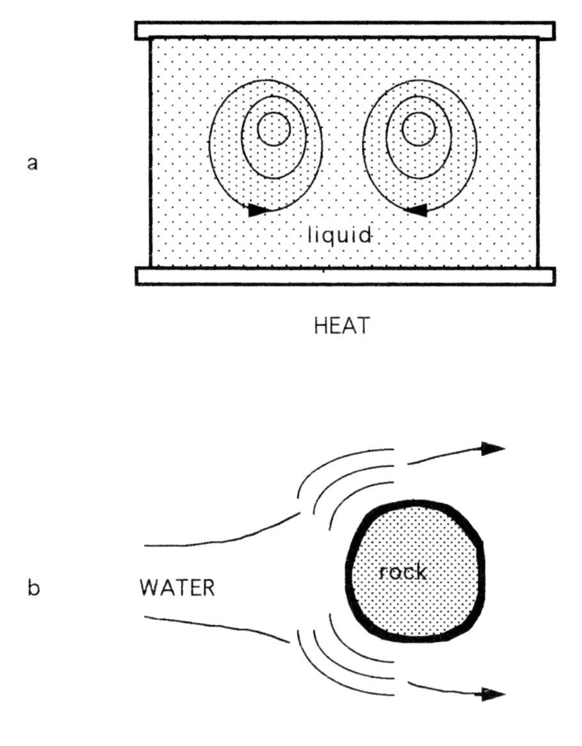

Figure 3. Flow patterns

planetary formation and so on. All have conspired to produce life as we know it. From various standpoints, life appears to be a natural outcome of the existence of matter. Of considerable interest, therefore, is the possibility of life existing elsewhere in the universe. Indeed, a programme is in place for the Search for Extraterrestrial Intelligence (SETI), involving the scanning of radio signals for features indicative of intelligent beings. It may be supposed, given a strong inherent tendency for matter to order, and the likelihood of suns with planetary systems akin to ours, that life permeates the universe. But the details are likely to be significantly different from what we know, since initial conditions and environmental factors are so critical for outcomes.

IS THERE AN UNDERLYING DRIVING FORCE FOR ORDERING?

As 'ordering' is such a universal phenomenon, attempts have been made to demonstrate an underlying driving force for it. One general approach, identified by Schrodinger in his seminal book, 'What is Life?', is to use the concept of entropy (the thermodynamic measure of disorder). According to the second law of thermodynamics the entropy of an isolated system (one not interacting with its surroundings) will increase to a maximum at equilibrium. Thus, for the universe as a whole, entropy will be increasing. However, this does not preclude spontaneous ordering occurring in isolated regions. It is possible to imagine negative entropy (order) being squeezed into such regions, or entropy excluded from them, at the expense of an increase in entropy in the environment (Fig. 4). None-the-less, this does not identify a particular driving force for the effect.

More recently Prigogine and coworkers have described ordering in terms of what are called 'dissipative structures'. These relate to the concept of energy/matter flows being inhibited, referred to above. The inhibition amounts to non-equilibrium behavior and dissipation refers to a particular description of energy change. In the most general situation of the universe, the energy of the 'Big Bang' may be taken as the source of these flows. While this appears to identify a fundamental driving force, questions remain as to whether the approach provides a completely adequate explanation of ordering, especially in relation to evolution.

Throughout this article attention will be focussed on two particular features considered to be significant in the development of 'dissipative' order. One is the universal capacity of units of matter to stick together or condense into stable entities; the other is the particular character of energy dissipation. Both are seen to be consequences of the 'Big Bang' event. The former relates to the binding forces of nature, involving at the chemical level, the different kinds and magnitudes of chemical forces. The latter is an aspect of the universal tendency for heat energy to be dissipated to the environment. Once the influence of these forces is assessed within the overall context of energy flow, ordering may be seen to be less mysterious. However, this should not diminish, but rather enhance appreciation of order in our universe.

Attention is initially focussed on spatial ordering. Comparable features are then identified for temporal ordering, thereby providing a basis for understanding more complex systems containing both kinds of order. The article concludes with a brief consideration of chemical and biological evolution in terms of the concepts of

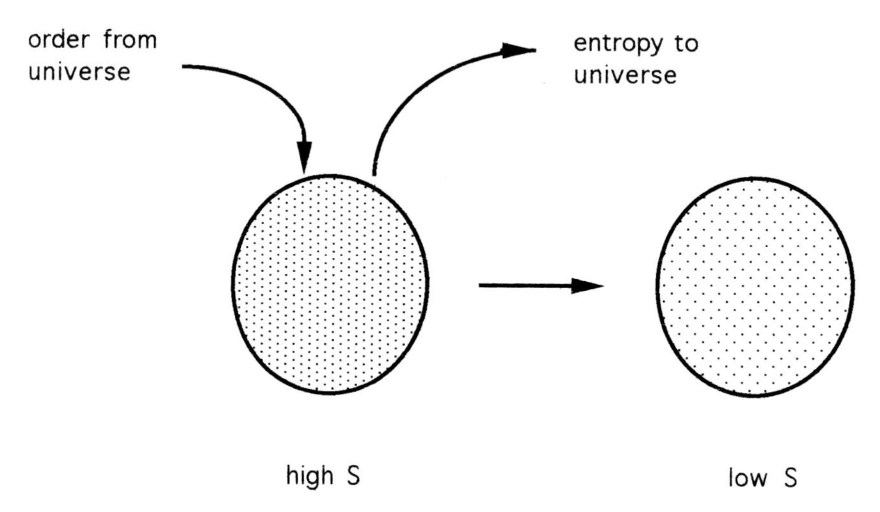

Figure 4. Localised ordering in universe

ordering developed. Of particular note throughout is the role in ordering of intermolecular forces (such as hydrogen bonding and van der Waals forces) facilitating the incorporation of environmental influences.

THE MEASURE OF ENTROPY

Most approaches to ordering are based on the concept of entropy (S), a measure of randomness in a system. In general the lower the entropy content of a system the higher is its state of ordering. Understanding entropy in a thermodynamic sense requires consideration of the origin of contributing factors. For a monomolecular system (eg., H_2) contributions come from the possible dynamic motions of the molecular assemblage. These fall into the categories of translational, rotational and vibrational movements. Each possible mode corresponds to an internal, quantised energy state. The total entropy is then derived from the summation of all possible states occupied at a particular temperature. The overall motion of a single molecule may be shown to correspond to a well-defined and fixed 3N 'ground state' modes or microstates (where $N =$ the number of atoms). The translational motion may be described in terms of 3, as also is the case for rotational motion (except that this is 2 for a linear molecule). The remainder, $3N-6$ (or $3N-5$), correspond to the possible vibrational motions. Thus, the greater the number of atoms a molecule has the greater the amount of 'entropy' energy it can hold or contain.

Such an analysis provides a firm basis for understanding entropy. It may be extended to cover multiple molecular assemblages whether gaseous, liquid or solid. Any complex system will have a range of microstates of motion, amplified (for more complex systems) by a range of possible spatial configurations. A certain amount of energy may be associated with each state, with the total entropy contribution arising from a summation of these. The larger the number of states the larger, generally, this value. A system becomes more ordered when the range of possible state combinations decreases. The system has become restricted in some way from expressing a more random (larger states) kind of behavior.

The entropy of a monoatomic gas, such as argon, is solely due to translational motion. Since its atoms will be moving more rapidly at higher temperatures (corresponding to occupancy of exited-state translational energy levels) the entropy energy increases with temperature. The system will then hold or contain more energy than would otherwise be the case. The same situation applies to the other forms of motion. Thus the 'entropy' energy content of any system is determined by the temperature of the system and the number of microstates available to it ($T \times S$). This constitutes part of the 'free' energy (G) of a system. The other component, the enthalpy content (H), is determined by the total energy contained in the constituent chemical bonds (a negative quantity). The total G (negative) is obtained from $G = H - TS$, so that for S positive, the situation is as shown in Fig. 5a.

The central driving force for change in nature is the dissipation of this free energy (the energy content of a system), according to dG (G products $-$ G reactants) $= dH - TdS$. Reactant chemicals in an energy state or well will transform into a more stable, product energy well if dG is negative, provided the energy barrier

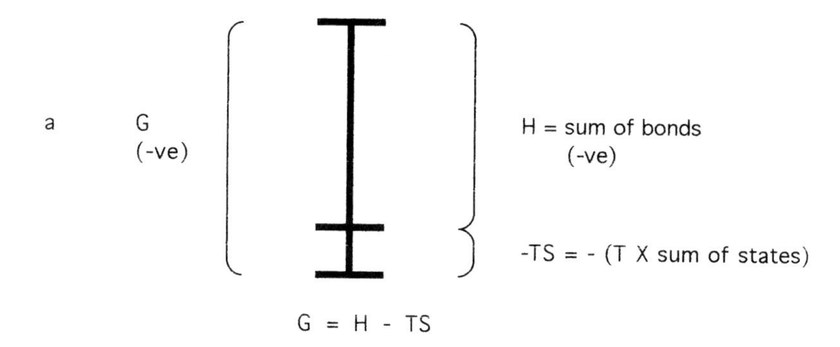

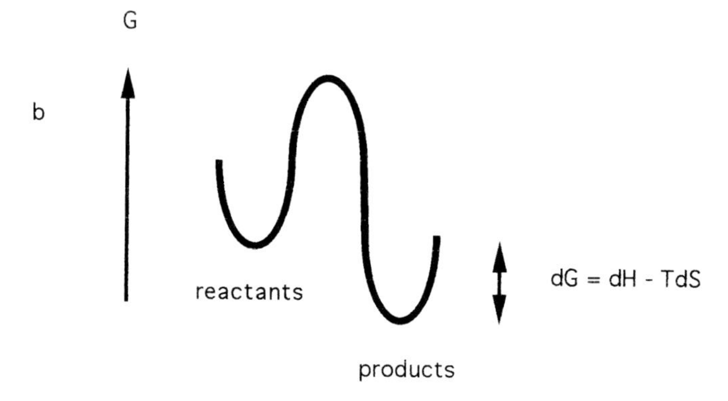

Figure 5. Energy relationships

is not too high (Fig. 5b). Dissipation occurs in two main ways – loss of heat energy to the environment during bond breaking and remaking, and spreading of energy into entropy states (i.e., increasing S). In some cases the latter may be sufficiently high to permit weaker bonds to form, so that dH is positive and the reaction endothermic. More generally, reactions are primarily driven by an overall increase in bond strength for product entities (making dH negative and the reaction exothermic). The greater the associated heat dissipation, the greater is the driving force for the reaction (with dH being directly related to dG). This is enhanced by dissipation of energy to entropy states (due to dS becoming more positive and the -TdS term more negative, making dG more negative). Thus, for example, reactions producing gases are favoured by the additional translational ('entropy') energy involved.

ENSEMBLE VARIATION

Features of product entities are generally of greatest interest in any analysis of the development of order. The reason is that ordering, as we perceive it, is normally the formation of a more ordered product than might otherwise have been anticipated. Thus, an understanding of order generally amounts to consideration of one product form with respect to other possibilities (as in Fig. 1a). This cannot arise for situations, especially gases, where the system can only exist as a single ensemble. However, once molecular complexity of a system is sufficiently high, different possible spatial ensembles may exist. This is primarily due to the special magnitude of intermolecular forces in relation to normal thermal and environmental influences. It is in these situations that a range of possibilities arises, where a more ordered ensemble becomes distinguishable from a less ordered one. In turn, this raises the question as to which state becomes favoured by a particular set of conditions. The example of the formation of a glass from a molten liquid may be used to identify what is involved.

Normal glass contains molecular units interlinked in a complex and disordered manner. But, alternate more-ordered states are possible when component entities become aligned to a greater extent. Indeed, over a long period of time such alignment leads to the glass becoming crystalline. In this case the original glass was trapped in a disordered state by kinetic factors. There was not enough time available for the molecular units to properly align, but with time alignment occurs (Fig. 6).

One way to describe the crystallisation of glass is to imagine that normal thermal motions facilitate the leakage of stored 'entropy' energy from the system. This would lead to a decrease in entropy content which, in turn, may be correlated with an increase in the state of order of the system. The entropy content would be higher for a glass since there would be more configurational states available than for the crystalline situation. The more the configurational variation within a solid, the greater will be, say, the number of distinguishably different lattice vibrational states (contributing to the entropy content). However, such entropy leakage could not occur to any extent by itself, since that would correspond to a positive change in G. None-the-less, it could happen if accompanied by a concomitant change in H.

While the strong internal bonds (and the enthalpy terms associated with them) would remain the same, different packing interactions would lead to different overall enthalpy values for the glass and crystallised-glass states. The more ordered state would optimise weaker intermolecular bonding, leading to dH becoming more negative. Consider four spherical entities (A) surrounding a fifth different one (B)

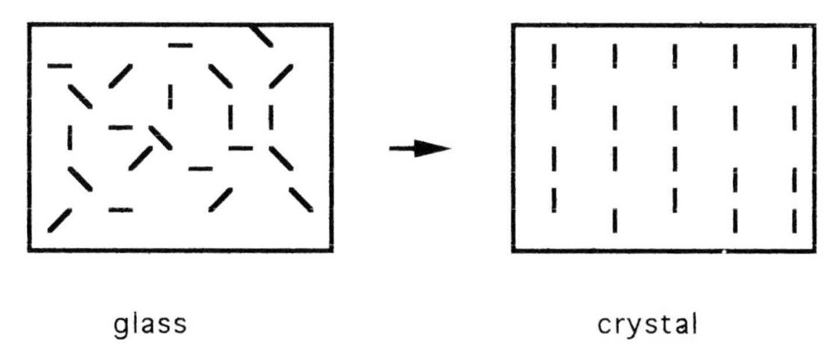

Figure 6. Ordering of glass

in a plane, as for Fig. 1a. There will be an optimal distance A-B where the van der Waals interactions are maximised. This in itself favours a symmetrical arrangement. Unsymmetrical ones will give a less-negative enthalpy content since B entities placed at closer or longer distances than A-B will make the system less stable. This process may be described as 'enthalpy leakage'. Consequently, ordering could involve complementary entropy and enthalpy energy leakage, or dissipation, to the environment.

ENTHALPY-ENTROPY DISSIPATION

The foregoing relates to the central feature of the Prigogine 'dissipative structure' approach to ordering, the concept of entropy dissipation. The latter may be considered to be the net effect of internal change of a system and external exchange with surroundings. But, as just considered, this is likely to involve simultaneous enthalpy changes within the system. Moreover, such a combined process could occur at constant free energy content, G (Fig. 7). Entropy energy dissipation leads to S becoming less positive while enthalpy dissipation leads to H becoming more negative. Since the H effect would be a second-order one (being related to intermolecular bonding) it is likely to be of the same order as TS, thus counterbalancing or compensating the S change to allow G to remain constant. Indeed, this would be a favourable situation since, as mentioned, an overall positive change in G (corresponding to a decrease in S only) would not be permitted. Also, any capacity for negative G change (i.e., better binding) is unlikely to occur without a decrease in S. Due to its distinctive character, energy change of this complementary kind will be referred to as 'enthalpy-entropy dissipation'. Such a concept provides a more complete thermodynamic approach to ordering.

A comparable 'enthalpy-entropy compensation' effect has been observed for reactions such as those of homologous series of organic compounds. This effect appears to be related to solvation influences on S and H of the solvent, water. Thus, for a large R (organic) side-chain, H may be decreased, negatively, by greater repulsion effects while S is increased by an increase in water structure disruption (disorder). The opposite would be the case for a small side-chain. Consequently dG

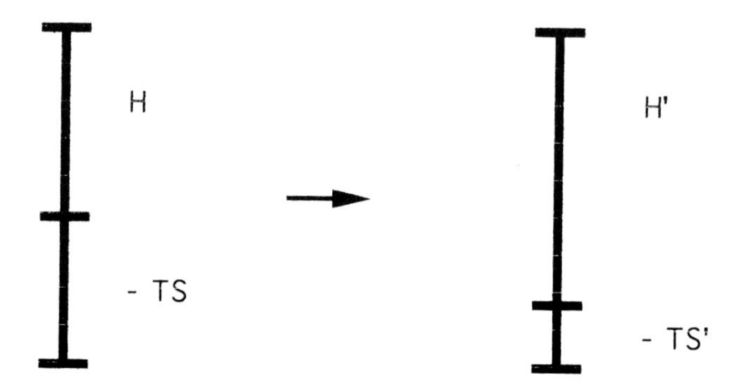

Figure 7. H/TS dissipation at constant G

would be dominated by the reaction of the functional group while second-order solvation effects would be adjusted in a compensating manner (due, presumably, to the variability possible for hydrogen bonding in solvent water). That this situation is similar to the ordering one, in terms of compensation between H and S, suggests the two have the same basis.

Another important feature of constant-G dissipation is that the activation barrier for change would be low, and therefore attainable via normal thermal agitation (as for the example of the crystallisation of glass). The process may be regarded as a second order one with respect to normal free energy chemical-reaction change. It involves the optimisation of weak intermolecular forces and concomitant loss of 'enthalpy-entropy' energy. Consequently, for a fixed molecular ensemble, this may be identified as a fundamental driving force for ordering. It corresponds to the normal ordering effect of strong bond formation, as considered below in relation to 'stickiness'. The effective difference for weak bonding is that it is not fully developed for many ensembles, because of the competing effect of disorder. As for normal reaction processes, this driving force may be amplified by energy or mass (mass action) input. However, it may be inhibited by randomising influences such as elevated temperatures and kinetic and/or intermolecular factors counteracting the ordering of component entities.

DISSIPATIVE STRUCTURES AND BIFURCATIONS

The Prigogine approach describes ordered arrangements as 'dissipative structures' and uses the concept of 'bifurcation' to identify how these arise. Bifurcations develop from instabilities, situations in ordering systems where outcomes are ill-defined. When a system is progressively removed from equilibrium, a bifurcation point may be reached, beyond which two alternate pathways exist (Fig. 8). The one that the system adopts depends critically on precise conditions, governed by either internal fluctuations or external influences. The two alternate pathways correspond to dissipative structures that arise from 'enthalpy-entropy dissipation'. Bifurcation enables a system to irreversibly shift to a new energy configuration from which it cannot readily revert to a former state.

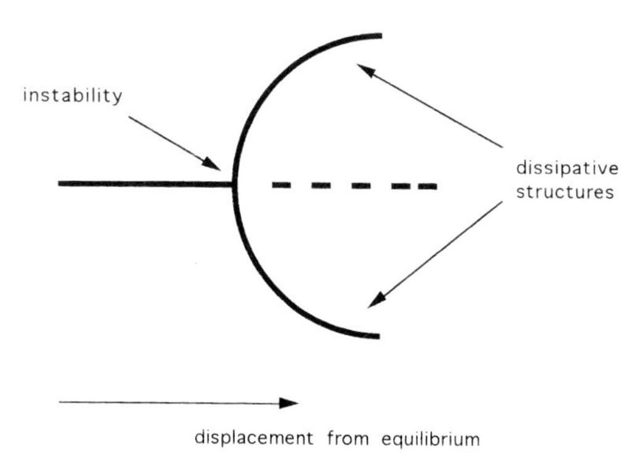

Figure 8. Bifurcation set-up

'Enthalpy-entropy dissipation' is an inevitable outcome, given the right balance of intermolecular binding and sufficient time. It involves participation of the environment to a greater or lesser extent, thereby inducing instability and the possibility of bifurcation leading to ordered states. Thus 'enthalpy-entropy dissipation' is the formative driving force for ordering phenomena. It is closely related to traditional free energy thermodynamic emphases that embody the overriding tendency for energy to disperse as heat and to spread to entropy states, during chemical transformation. 'Enthalpy-entropy dissipation' is a comparable universal tendency, being residual energy of any system that leaks out, in an alternate way, to the environment. It involves the same optimisation of bonding interactions, but shows up as a distinguishably different effect by being restricted in extent by disordering influences. Such extended and generalised thermodynamics may be applied to all chemical systems, whether equilibrium or non-equilibrium.

STICKINESS OF MOLECULAR ENTITIES

As already mentioned, a key feature in complex systems is the way in which component units associate or stick to each other. Indeed, 'stickiness' of entities may be viewed as the universal determinant for all ordering processes. While this is more readily appreciated for spatial ordering, it may also be extended to include temporal ordering, as considered later. The operative solid state process is spatial packing that leads to symmetrical ensembles (Fig. 1). When insufficient time is available for the most symmetrical arrangement to form, ensembles may become trapped in lower-symmetry states (as in the case of glass formation).

The concept of 'stickiness' may be applied to all steps of matter formation arising from the 'Big Bang' event. Fundamental particles came together to form atoms, atoms linked to form molecules and so on. At the level considered here, molecular entities associate and condense and reaction pathways interlock. Looked at in this more general way, 'stickiness' is essentially due to the quantum mechanical effects of atom and molecular formation, involving the strong, weak and electrostatic forces. These may be considered to have arisen from the huge energy of the 'Big Bang' not being able to be kinetically dissipated in a smooth manner. Resistance gave rise to binding effects and these caused entities to coalesce rather than to fly apart on colliding. Nucleosynthesis in suns and molecular formation in interstellar gas clouds are examples of ordering involving these forces. The force of gravity operates on a grander scale in a comparable manner. All may be regarded as order-producing situations involving the binding of components of matter together by the forces of nature. However, it is normally the formation of ordered molecular assemblages and integrated reaction schemes, leading to the molecular development of life, which are the focus of attention in discussions of ordering.

The latter are characterised by weak intermolecular (or interionic) interactions, such as hydrogen bonding and van der Waals forces. These become particularly significant due to a marked susceptibility to longer-range environmental influences. This can produce a range of slightly different situations, as indicated in Figure 2. At one extreme 'stickiness' may be so strong that the system becomes stuck in one fortuitous form, while at the other entities may dissociate completely into just one possible configuration, as for a monoatomic gas. However, in-between levels of stickiness allow a wide array of ordered states to arise as a result of varying environmental influences.

Olmsted, in "Entropy, Information and Evolution", has similarly identified forces as the source of ordering. But, in relation to a comment about Coulombic bonding being the source of chemical (entity) ordering, he has stated: "The assembly of prebiotic molecular fragments into more and more complex sequences . . . does not appear to be a similar slide into a potential energy well". However, weak inter-unit binding enables the same force-ordering process to be continued at higher levels of organisation. Yet the effect is more subtle because of disordering influences. As explained later, this allows evolution to follow the axis of greatest richness of ordered possibilities, corresponding to maximum effective contact with the environment. Although the latter may appear to be a 'random' influence, it constitutes information and as such informs the evolutionary processes. Furthermore, because of the special nature of 'weak bonding' ordering, the overall thermodynamic situation is also unique. This highlights the particular significance of 'enthalpy-entropy dissipation' as a driving force counterbalancing tendencies towards disorder.

The concept of 'stickiness' has been used to explain the abnormal five-fold symmetry feature of quasicrystals. This is based on Penrose's 'tiling' concept, involving the fitting together of tile-like entities. By invoking a variation in the stickiness of the tiles' edges, a satisfactory explanation of the observations may be obtained. This is a good example of the subtlety of spatial ordering. Later, the same concept of 'stickiness' is applied to temporal ordering, where reactions coalesce to give integrated, rather than dispersed outcomes.

THE FEATURE OF TIME

Ordering of the inter-unit kind corresponds to varying degrees of rearrangement of a basic set of molecular entities. Manipulation of a kaleidoscope may be used as an illustration of what is involved. Juggling it randomises the pieces against the tendency for them to 'stick', or fit together, i.e. to order. The combination of these two effects produces a range of different patterns, the detail of any one being dependent on the nature of the energy flow (juggling process). This is a function of both extent and time of the disruption. For certain chemical systems the shape of entities may lead to an additional time feature. Beyond a certain length limit, linear molecules may become so entangled that very long times are required for disentanglement. The system becomes kinetically trapped in an adventitious form, as in the case of glass formation.

Also, as mentioned, the magnitude of flows plays an important role in the attainment of order. This relates to energy input, electrochemical or photochemical, and mass flow (mass action) effects. The greater the input the faster the process involved. These effects are generally of greater significance for temporal ordering, considered below. Another important factor for the latter is the timing (rates) of chemical reactions. These determine the overall dynamics of complex reacting systems in a crucial manner, because of a mutual interplay with diffusional features. A fundamental requirement for viable life processes was the speeding up of essential metabolic reactions, as achieved by enzymes. Adjusting rates to fit in with other processes in an ordered manner would also have been a key requirement. Again, it would have been the inherent ability to finely tune rates, via changes to enzyme structure, which would have been important. Time plays a crucial role in determining the nature of both spatial and temporal ordering.

SELF-ASSEMBLY

A ubiquitous chemical phenomenon, akin to crystallisation, is self-assembly of molecules in aqueous and other media, to produce ordered assemblages. This invariably involves the clustering of hydrophobic centres leaving hydrophilic ones to interact with an aqueous environment. Surfactants form micelles of varying spatial shapes (Fig. 9a), lipids aggregate to form liposomes, while many proteins form globular shapes via such inter-molecular binding effects. Many biological structures, such as cell walls and microtubules, form from component building blocks in the same manner. A critical factor in terms of versatility of function is a balance between rigidity (order) and flexibility (disorder) that allows a wide range of possibilities to accrue from a single basic arrangement. Aggregate ordering is essentially based on

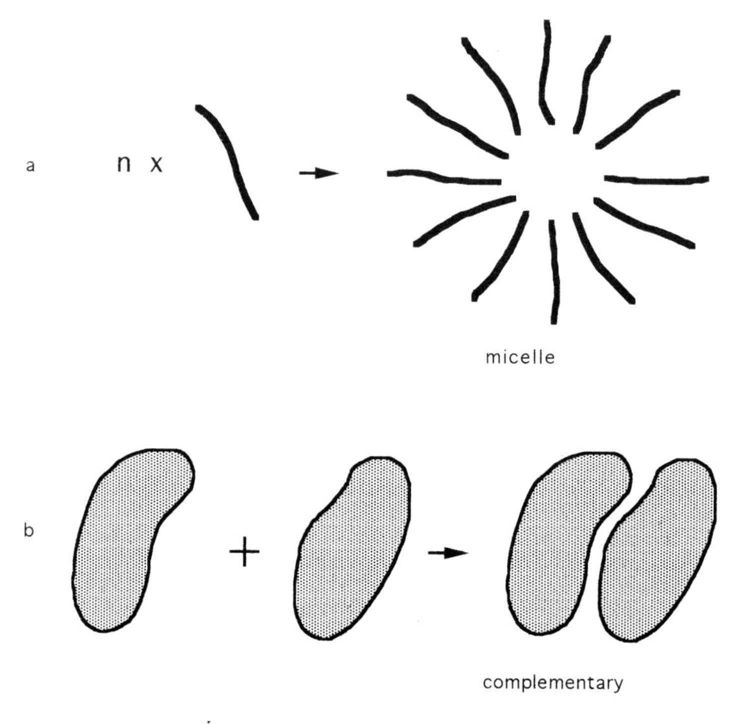

Figure 9. Molecular association

the sticking of molecular entities with each other in optimal geometrical ways. A subtle, but highly significant variant is complementary association, where differently shaped subunits interlock (Fig. 9b). This occurs when single strands of DNA bind to give duplexes. Bases stick together in a pairwise manner with the A-T and G-C pairings having the same overall lateral dimensions. The versatility of such arrangements is also related to the weak nature of the inter-unit forces involved.

CRYSTALLISATION ORDER

Crystallisation is a key process to consider in relation to ideas about ordering. Here the level of ordering is generally extensive. The normal process of crystallisation may be regarded as a situation where the laying down of new material on a growing surface is generally ideal for 'enthalpy-entropy dissipation'. Under such conditions, sufficient time and disruption are available for the juggling of entities into the most symmetrical state. This allows initially formed imperfections to be annealed out, with the levels of 'stickiness' being sufficiently strong to ensure ordered arrays are readily retained (a in Fig. 10). Indeed, for the ionic and covalent classes of crystals, strongly directional inter-building-unit forces lead to well-ordered crystalline products. For the molecular and intermediate, ionic-covalent crystal classes more time and annealing may be required for the production of well-formed crystals.

Disorder in crystal lattices is quite common. This is an indication of the competing tendencies for order and disorder. An important class of crystal is that giving rise to patterned external forms, referred to as dendrites. The phenomenon is a general one, especially for molecular crystals, where the intermolecular forces are relatively weak. In these cases the crucial feature is the rate of crystallisation, where a balance between a tendency for ordered growth, and disruption of that order, occurs. Neither a well-formed crystal nor an amorphous product is formed, but an intermediate type.

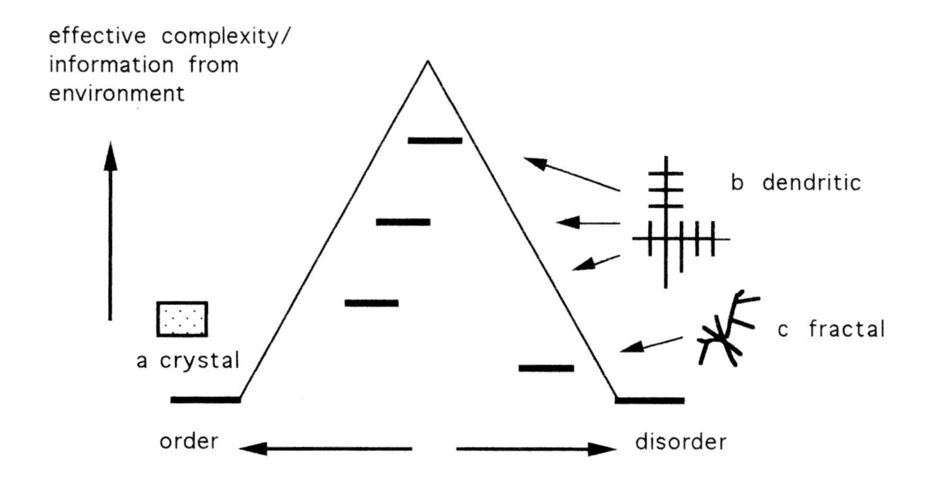

Figure 10. Crystal growth variation

This is characterised by patterning features of the external form of the crystals (as indicated by b, Fig. 10). An additional level of symmetry develops under these non-equilibrium conditions and the pattern formed is a reflection of the external conditions involved. Depending on precise conditions, a range of different dendritic forms may be obtained for a single system (Fig. 10b). This effect may be related to a more general feature of crystal growth.

In any particular crystallisation of a compound most of the crystals formed on any one occasion invariably have a similar external appearance, especially if weak inter-unit bonding is involved. Under different conditions the same compound may yield crystals having a distinctly different basic external form. This highlights the specificity of environmental influences. While correlation with particular long-range effects has not been established, environmental detail is in principle decipherable from such growth features. The particular morphology of each individual snowflake reflects the global history of its constituent water molecules. This arises from the interplay between 'enthalpy-entropy dissipation' and counteracting disorder. Pattern type also depends on the nature of the mass/energy flow conditions employed, as indicated by the kaleidoscope illustration, mentioned above. An empirical relationship has been shown to exist between entropy dissipation, order and driving forces, for dendritic growth.

The foregoing enables an important correlation between complexity and information to be made. At the high order extreme (a, Fig. 10) entropy content is low and essentially a function of degrees-of-freedom microstates only. At intermediate levels (b, Fig. 10), where external form varies according to environmental influence, the number of states is higher. These additional states correspond to macroscopic features and correlate with information entrapped from the environment. Further towards the disorder side, corresponding to fractal growth (c, Fig. 10), extra macroscopic states accrue, due to the existence of additional growth configurations. But, in practice, these are indistinguishable, making complexity low. Information content is also low because the growth patterns effectively arise from a single environmental influence. Thus, the position of maximum complexity corresponds to the maximum effective attainment of information from the environment (Fig. 10).

FRACTAL GROWTH

At the higher levels of environmental influence, where disorder leads to fractal growth of the frost pattern kind, patterning is less readily discernible. However, while such products are amorphous in terms of poor X-ray diffraction, a special type of ordering persists. This arises from long-range correlation effects determining how minute component particles stick together. Analysis of the resultant patterns shows regularities that are scale independent; the same basic shape appears at different levels of magnification. This remarkable effect may be referred to as scaling order to distinguish it from dendritic or symmetry ordering. It probably reflects the widespread existence of fractal features embedded in the matter world. As such, fractal growth represents encapsulation of a single, gross informational feature of the environment, rather than a range of variable features, as for dendritic growth.

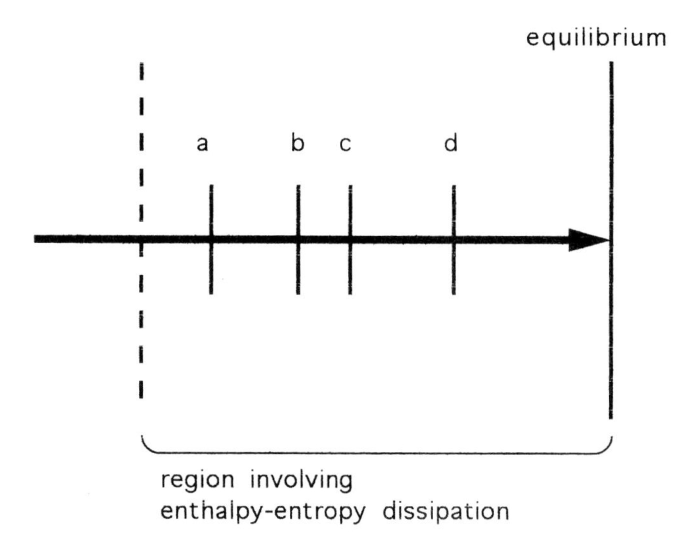

Figure 11. Ordered states, a-d, in non-equilibrium domain

To summarise thus far, ordered structures may arise under particular conditions that can now be well defined. Essentially, ordering is the consequence of two interrelated features of matter: (1) the quantum mechanical one of binding (the stickiness of matter) and (2) the universal tendency for free energy to be dissipated to the environment. Largely due to an emphasis on the tendency for entropy to increase, an impression has developed that ordering is exceptional. But it is a natural outcome, in the case of chemical ordering, of ionic and covalent bond formation. However, in the case of molecular ensembles involving weaker intermolecular binding, environmental influences become incorporated into the ordering process. This produces a range of ordered states, via 'enthalpy-entropy dissipation', in the non-equilibrium domain (Fig. 11). The richness of this resultant region of in-between order has far-reaching implications for the processes of chemical and biological evolution. At this stage the discussion switches from spatial to temporal ordering as a prelude to considering evolutionary development.

ORDERING OF CHEMICAL REACTIONS

So far the emphasis has been largely on the solid state because this offers a more straightforward introduction to ordering, of the kind being highlighted. While physical ordering can occur for the liquid state (as in Benard structures) in a comparable manner, the level of complexity increases profoundly when chemical reactions are involved. Such kinetic ordering has been considered from various viewpoints and in great detail. Turing was the first to recognise that a combination of reactive and diffusional conditions could lead to temporal ordering. Field, in highlighting that the only essential requirement for reaction ordering is a non-linear

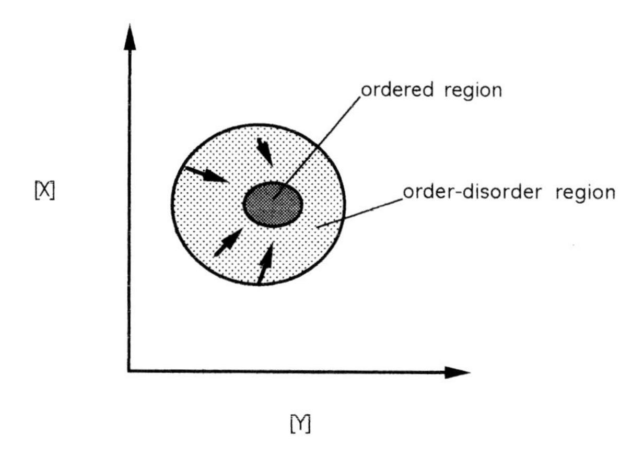

Figure 12. Evolution of system to ordered state in XY phase space

kinetic feature, has commented that "such terms occur in essentially all chemical reactions of any complexity especially in biochemistry". This essential feature of non-linearity may be loosely correlated with the concept of 'stickiness' used in solid-state situations. Such effects restrain reacting systems from proceeding to equilibrium. Prigogine and coworkers have applied bifurcation theory to particular reactions in further characterising dynamical ordering processes. Moreover, it has been discovered that certain systems can exhibit the highly-ordered dynamic behavior of chemical oscillation.

Also of interest is the manner by which a system may evolve from a range of situations to a state of kinetic order. This can be regarded as a reflection of the influence of 'enthalpy-entropy dissipation' operating on the reacting molecular ensemble. While the fine details are complex, any non-linear feature is likely to involve weak-bonding association. This relates to the formation of multi-component intermediate species, as in catalysis (a key feature of kinetic ordering). Of necessity, the bonding in these species must be relatively weak in order to permit dissociation to occur. Consequently, the character of these species will be affected by just the same influences considered for the solid state. Inter-unit bonding will be subject to environmental effects, leading to long range correlations. A balance between a tendency for ordering ('enthalpy-entropy dissipation') and one of disorder (dispersion of reacting species) will again operate. This means a system in reaction – diffusional flux may evolve from a range of positions, in what is called phase space, to regions of defined order (Fig. 12). In each case the transition would be driven by enthalpy-entropy dissipation' and be influenced by particular flow characteristics of the system. Such a description relates closely to the concepts of fluctuation, instability and bifurcation.

INSTABILITIES

An essential feature of all non-linear systems is the existence of instabilities or bifurcation points, considered above. These allow for environmental factors to have a marked influence on future development. In general, such delicate balance points provide the widest range of outcomes. The manoeuvrability of birds in flight derives from an aerodynamic state of instability. The variability of dendritic snowflake form is a solid-state analogue. In the case of complex reaction systems, compositions may be identified where there is a marked sensitivity to environmental effects (referred to as 'noise' in this field). In all of these cases the environment informs the behavior of the system, due to the latter's state of instability. The system incorporates information from the environment in a profound manner, even though that information is normally inaccessible to us. However, we may recognise this happening by particular features of temporal ordering, such as steady state behavior, which correspond to negative changes in entropy. As found for the solid state, these are normally subtle in character. An extension of this analysis may be made to the linkage of one system to another. Information may be passed from one to the other, yielding a richer overall system in terms of viability and versatility (considered in more detail below).

STEADY STATES

Another context within which to consider temporal ordering is the non-equilibrium attainment of two (bistability) or more metastable states for reacting mixtures. These correspond to different overall modes of integrated reaction for the same composition. The key requirement is non-linearity of the kind usually associated with autocatalytic feedback processes. In effect, the overall reaction process is held up from proceeding to equilibrium by steady state (pseudo equilibrium) formation. Again, as for the solid state, this behavior only arises beyond a certain minimal level of complexity. The reaction conditions required usually have to be of the 'open'/'continuous' rather than 'closed'/'batch' kind (ie 'closed' in the matter, rather than

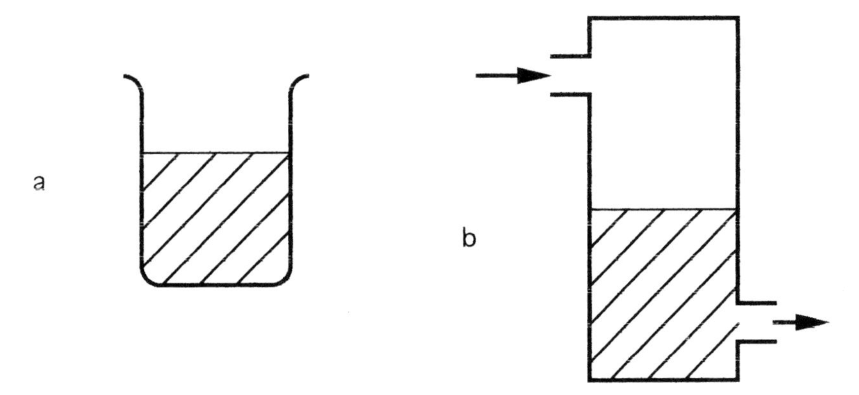

Figure 13. Batch and CSTR reaction systems

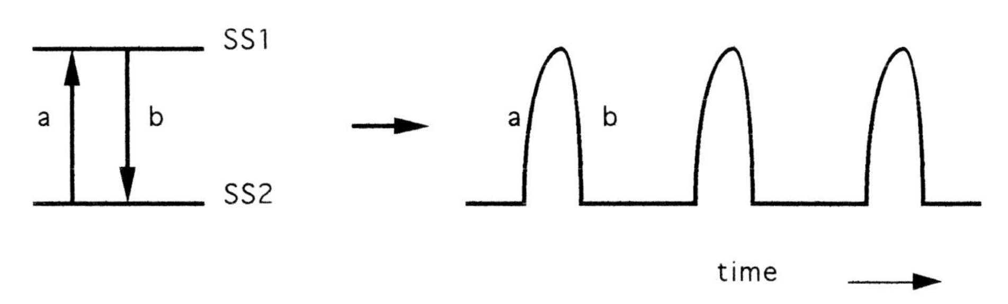

Figure 14. Oscillation between steady states

energy, sense). The former is attained by having reactants flow into a vessel (CSTR-continuously stirred tank reactor, Fig. 13) to displace already reacting solution. By adjusting flows appropriately one or another distinguishable steady state may be attained. Thus, for one composition and one fixed set of reactions, two or more states may be realised by varying flow rates (which, in turn, influence diffusional rates). Ordering is again a consequence of order (linked chemical reactions) being counter-balanced by a tendency for disorder (diffusional dispersion). Alternatively, it is flow enhancing 'stickiness' of individual reactions against a tendency for reacting entities to disperse. Such ordering is probably universal in biological systems, although difficult to identify. Much more recognisable and dramatic is chemical oscillation, where a system flips periodically between steady states, due to the availability of switching mechanisms (Fig. 14).

CHEMICAL OSCILLATION

The development of oscillatory behavior is a case of chemical ordering akin to high quality crystal formation. The specificity of the reaction manifold (the set of reactions interlinked to ensure cyclical behavior) plus appropriate diffusional parameters (comparable to the 'juggling' feature) ensure 'automatic' ordering. When the system occupies such a well-defined region of oscillatory phase space it is relatively unaffected by environmental factors. Other, less-precisely interconnected reaction schemes will show greater sensitivity (as for dendritic crystals having weaker inter-unit binding). But these display additional features of ordering (as for the solid-state analogues), such as frequency-doubling and intermittent oscillations ('chaotic' behavior). Features of this kind may be studied by purposely displacing 'good' oscillators from oscillatory regions of phase space. As for spatial ordering, matter/energy flow characteristics are crucial in determining fine details of such ordering. Certain systems, that may not oscillate under batch conditions, will readily do so under flow conditions, since flow increases the possibility of forcing a system into an ordered state.

The most celebrated example of chemical oscillation is the Belousov-Zhabotinskii (B-Z) reaction, the acid catalysed bromination of organic acids. This involves about 20 separate reactions linked in an integrated manner. Oscillatory behavior is readily attained under stirred batch conditions for a wide range of concentrations. When a redox indicator like ferroin is added, the oscillations show up as prominent periodic changes in colour. Many other chemical oscillators are now known, although most are only achievable under flow, CSTR conditions. Oscillatory behavior has also been established for various biological systems, notably the glycolytic one.

IRREVERSIBILITY

The development of order is invariably associated with the elusive phenomenon of irreversibility. This may be considered in relation to 'enthalpy-entropy dissipation' and Coveney's comment, ". . . irreversibility seems to emerge as a necessary consequence of the instabilities in more complicated systems." Instabilities represent the influence of long-range effects on outcomes. As a consequence, exterior or global influences become historically incorporated into the system. This happens because the system evolves into a state of order characterised by local order being balanced by 'disordering' influences from the environment. 'Enthalpy-entropy dissipation' facilitates a system 'juggling' into an ordered state of this kind. As soon as such an environmental influence becomes part of the system microreversibility is lost. It is effectively overridden by the historical incorporation of the long-range input. The low probability of the system ever returning to the highly specific 'global' state it was previously in, makes the situation an irreversible one. Once order develops from an instability it spawns comparable ordering throughout the system, rather like crystallisation. An arrow of time may be identified with non-linear features incorporating information from the environment in an irreversible manner. As evolution develops along this axis information flow becomes increasingly more significant, culminating in the phenomenon of consciousness.

OPENNESS OF SYSTEMS

Closed (batch) systems usually proceed reasonably quickly to equilibrium with the build up and retention of product species, limiting the lifetime of ordered states like oscillatory behavior. By contrast, ordering of open, flow systems, including oscillatory behavior, may be retained indefinitely by maintaining necessary flow conditions. Interestingly, the living, cellular situation may be described as having intermediate open/closed character. A useful measure of openness for chemical oscillators is the lifetime of oscillatory behavior, which in turn is an indicator of the extent of overall ordering. Openness not only reveals the sustainability of a system, but it reflects a much more specific connectivity with the environment than so far considered. The influence is the precisely identifiable inflow of reactants or metabolites. Such features of 'openness' may be applied to all non-linear reacting systems, besides oscillatory ones. Since it is undoubtedly a crucial aspect of metabolic life-reaction pathways, the concept of 'openness' has a direct bearing on questions regarding a thermodynamic necessity for Darwinism. One study, taking rates and efficiencies of cell reactions as a measure of fitness, correlates well with this emphasis on openness as an indicator of chemical viability. The environmental feature of openness is thus of considerable value in analysing higher-level ordering as it provides a direct link between physical necessity and biological outcome.

ACCRETION OF ORDERED SYSTEMS

The next major step in analysing the development of order is that of assessing what happens when differently ordered systems are combined. This may be viewed as a key process in the overall evolution of life. In the broadest sense, energy from the 'Big Bang' has been encapsulated as matter entities in a series of ordering processes. At each stage of development new forms of matter emerge, along with new features of order, such as the symmetry of vibrational modes in the case of molecules. The combination of two or more ordered entities (in this case atoms) produces higher levels of order, offering new possibilities and outcomes. Thus, a multiplier effect operates, since the products are more than simply the sum of constituents.

For two ordered molecular ensembles linked together, multiplier effects will arise from the underlying thermodynamic influence of 'enthalpy-entropy dissipation'. Dissipation from one organised unit would 'inform' the organisation of another linked to it and vice versa. Each unit would 'feed' on the other. Such a concept relates to Schrodinger's view of systems becoming organised in lieu of 'negative entropy' being transferred to them from elsewhere in the universe. But, in this case the effect will be multiple, due to the two-way interaction. An even higher-level type of order combination operates for symbiotic association, discussed below.

The multiplier principle has been studied in relation to chemical oscillators. Temporal ordering of the B-Z oscillator has been linked to various spatially ordered micelle systems to produce new kinds of oscillatory behavior. For example, the period of oscillation becomes more constant and the lifetime of the oscillator is significantly enhanced, both indicating the attainment of a more open system. Other features support the idea that these binary systems develop spatial as well as temporal ordering. This is not so surprising, given that certain B-Z compositions, when not stirred, form spatial patterns. The micelle part of the binary system, although itself dynamic, would have the capacity to sequester B-Z components in a non-uniform manner. Such effects may be relevant to the maintenance of cell processes in living systems. Also, these may have been critical for the progressive evolution of chemical systems, leading to life. Thus, a change from micelle to liposome, as host for a dynamically ordered life-like system, could have been a key evolutionary step. Given the right conditions of chemical exchange between individual liposomes, an even longer-lasting, more open system would be expected. This progression in openness represents a continuum of influence of physico-chemical principles identified for simpler systems.

A key observation of binary B-Z systems is the extent to which these become more resilient to the addition of foreign entities. Thus chloride ions, normally incompatible with B-Z components, may be incorporated (presumably due to some sequestering effect). This points to an important selective advantage such 'richer', multiply-ordered, systems have.

ENCAPSULATION

The foregoing binary B-Z systems exemplify the significance of the phenomenon of encapsulation. Since the dynamic part is confined to a region of space by a spatial molecular shell, but not entirely so, these are effectively CSTR in character, thereby facilitating flow and openness. In chemical evolution, the advent of encapsulating, lipid bilayer, protocells would have enabled evolving chemical processes to be isolated and 'protected' from dispersion and 'outside' competition. For continuing development, this would have required a cell membrane with sufficient rigidity for encapsulation, yet enough permeability to permit matter/energy exchange. However, such versatility of spatial structure would have been assured, with evolution following the axis of maximum structural diversity.

Even temporal ordering is a kind of encapsulation. The linking of reactions into oscillatory cycles has the effect of constraining chemical systems against diffusional dispersion. Replication amounts to a comparable situation of containment of complex chemical behavior. Each case of dynamical or spatial encapsulation opens up a vast range of new possibilities for organised matter. While the emergence of higher-level features may seem dramatic and surprising to us, these none-the-less arise as a consequence of general ordering phenomena. Encapsulation multiplier effects also operate in the cyclical linkage of reaction pathways into 'hypercycles'. Extensive studies, both theoretical and experimental, have been made of such systems by Eigen and coworkers. These indicate that hypercyle 'encapsulation' would have been crucial for molecular evolution.

THE GROWTH OF COMPLEXITY

The net complexity in the matter world increases with the arrow of time following the order-disorder axis of maximum variability (Figs. 2 and 10). At each stage this balance ensures a maximum expression of ordered ensembles (Fig. 15). Due to the tight nexus established with the environment, these ordered states become appropriate and viable contenders for subsequent environmental change. Such an effect would operate at all stages of molecular and biological evolution. With variation being a built-in feature, the best-fitting entity would become the most viable (Fig. 15). There is no need to invoke 'selection' as a physical law; it remains no more than a tautology, as suggested by Popper. The evolutionary process is determined by specific features of spatial fitting and adventitious accretion arising from time to time. Such accretion, or 'stickiness', allows order to feed on order, thereby enhancing richness, viability and resilience. Yet, there is nothing sacrosanct about the particular details that emerge.

The deterministic time-line of our world was set in place by the initial conditions of the 'Big Bang'. But, as Gell-Mann has emphasised, increasing complexity follows a chequered course, characterised by very narrow windows of development leading to marked emergence of new possibilities (Fig. 15). However, the overall growth of complexity remains constant due to steady energy-matter flows, which also sustain existing states. New structures arise and persist due to the irreversibility of ordering

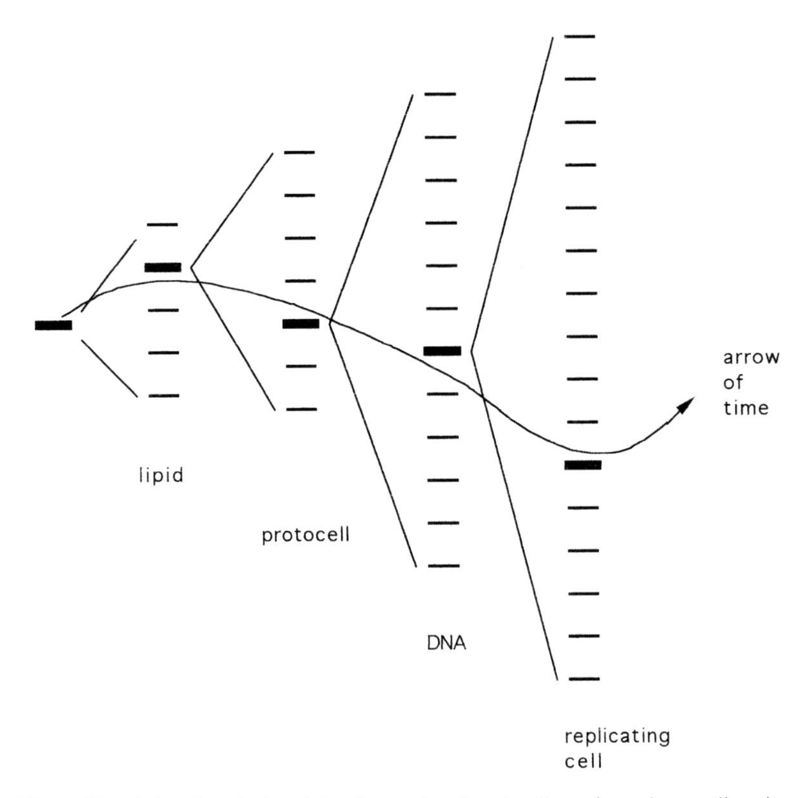

Figure 15. Axis of evolution following order-disorder line of maximum diversity and best-fitting emergent entities

processes and the appropriateness (in terms of environment) of outcomes. However, reversals may occur due to catastrophic environmental change and these may be local or planetary in extent. On a grander scale, some point in time will be reached when the potential of the 'Big Bang' runs out and the overall trend reverses. Regardless of our perspectives, we inhabit a highly deterministic world. Yet, by recognising and reflecting on this stark state, we may become free to independently determine consonant ideals.

THE ORDERING OF CHEMICAL EVOLUTION

A sketch of ordering steps likely to have been crucial for the evolution of life is now given. Of considerable significance is the origin of pre-biological molecules like amino acids. This was once thought to be terrestrial, but is now recognised to probably have been cosmic (Fig. 16). The key discovery was that interstellar gas clouds contain vast quantities of organic molecules being continually synthesised in space. Microwave signatures allow constituent molecular species to be clearly identified. While these species are dispersed very thinly, time intervals of extreme length ensured progressive chemical processing. To date about 80 kinds of polyatomic molecules have been found, ranging from simple diatomics to 11-carbon chains and biochemically important entities. Synthesis is enhanced by the adsorption of organic molecules onto dust particle surfaces, leading to the formation of larger molecules. Even in more diffuse interstellar regions, where radiation, unimpeded by dust particles, would be expected to destroy chemical species, other molecules both form and survive. New evidence suggests the presence of unsaturated carbon chain species and even exotic entities like fullerenes. Throughout space, atom combination, driven by radiative energy, fabricates the ordered building blocks of life.

Gravitational collapse of an interstellar gas cloud leads to the formation of solar systems. Much of the organic material is lost but some survives in the coldness of distant outer regions and, with accompanying dust, coalesces into 'dirty snowball' comets. On orbital passages around the sun many subsequently impact on planets (Fig. 16). Craters on our moon testify to the extent of this bombardment at earlier stages of solar-system evolution. The process continues, as viewed recently in the dramatic impact of fragments of the Shoemaker-Levy 9 comet with Jupiter. Much of the organic material is again pyrolised during impact, but some survives to become the starting point for further chemical processing. A planet having an appropriate atmosphere and radiative flux from the sun is an ideal host, especially if endowed with good levels of water from cometary impacts. Some chemical working could have already occurred within comets, where heating of outer layers during passages around the sun would produce liquid water. In all of these steps, ordered states would be combining and radiant energy utilised to ensure the production of progressively more complex chemical species.

Whether the first cell emerged in space or on planets is difficult to judge at present, but fossil evidence suggests life was present on planet earth less than 1 billion years after its formation. Furthermore, studies of moon rocks indicate the earth-moon system was in hostile activity for about half that time. Consequently, if only

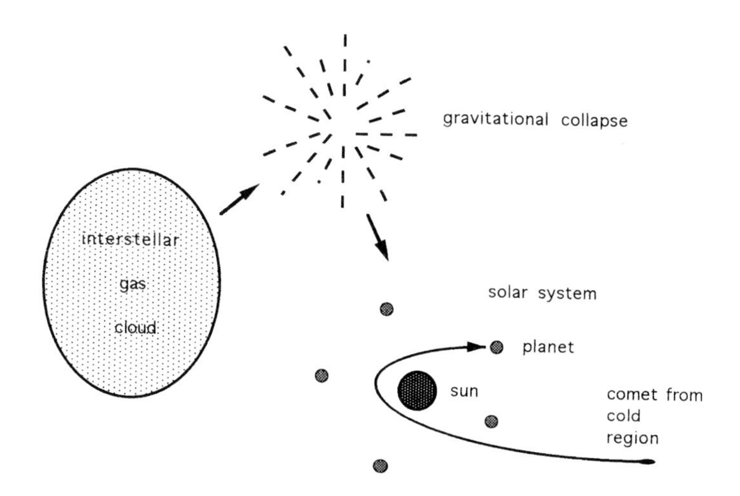

Figure 16. Traverse of life molecules from gas cloud to planet

precursor molecules arrived via comets, chemical evolution would have spanned a relatively short period of time. This may indirectly imply a higher level of extra-planetary development towards life. In any case, protection from degradation by excessive radiation would have been crucial throughout. However, this may be achieved, via variously 'ordered' locales, like the interiors of comets. For contemporary life on earth, the crucial protective feature is a thin layer of ozone.

As indicated above, the lifetime of ensuing chemical systems is an ongoing test of sustainability. A closed system eventually comes to an equilibrium state, normally in a relatively short period of time. Systems open to energy or matter flows are maintained in a steady state for considerably longer.

At some stage, protocells form to yield a new platform for chemical evolution. This spawns colonies of primitive cells with semipermeable walls, containing varying degrees of organised chemical content. These achieve high levels of openness, lifetime and viability. With a relatively constant planetary environment generations persist, through growth and decay of individual cells, for a long period of time. Such a matrix of variation and continuity allows gradual development of new features of organisation. Eventually replication takes over from simple replacement to endow primitive life with greatly enhanced survivability. Lifetime (in terms of sustained ordering) takes on a new meaning; it becomes the property of the species rather than of the individual. Moreover, this heralds a new capability to cope with environmental change. Emergence of the first replicating cell comes with just the right combination of cell structure, chemical combination and metabolism.

Key molecules like proteins and nucleic acids becoming sequestered into protocells would have facilitated the emergence of this primitive self-replicating cell. Various recent studies have identified likely chemical features of this process. A working model is available for cell (membrane) replication. Micelles (precursors to protocells) have been prepared that catalyse their own replication via autocatalytic processes. The term autopoiesis, from the Greek 'auto' (self) and 'poiesis' (formation), has been used to describe such systems. These are self-bounded and able to self-generate due to reactions taking place within their boundary. Molecular replication has also been demonstrated for a wide range of chemical systems, including DNA-like oligonucleotides 24 monomers long. The key factor for these is the geometrical fitting of one molecule with another via weak forces (H-bonding, van der Waals and aromatic stacking). If association has an appropriate degree of looseness it can endow on a molecular system greater survivability under altered chemical conditions. This exemplifies the versatility of the order-disorder balance mentioned above. In principle, such kinds of protocell and molecular replication could have combined to provide the basis for the development of total-cell replication. The results demonstrate that a wide range of plausible chemical processes, inimical to the development of life, would have been available at this stage.

Another associated development would have been that of coding and transcription. Although many studies and proposals have been made about this step, it remains to be more precisely elucidated. However, the general principles of molecular sticking and fitting, applied to nucleic acid-protein interactions, indicate how a primitive process could have originated. At each level, as just indicated, development would have followed the balance line of specificity and flexibility. These two inter-related requirements are satisfied by the particular characteristics of DNA, where accuracy of copying is counterbalanced by susceptibility to change, via mutation. During these developments proteins would also have taken on an ever increasing role as master molecules. Stringent requirements for overall integration of chemical behavior would have required the tight control of metabolic and catalytic processes. Proteins became enzymes by accreting entities like transition-metal ions and by evolving active sites, via fine tuning of side-chain character. Such polypeptides speed up key processes to enable time-wise compatibility of the reactions of life to develop.

Once the first self-replicating cell is formed further association leads to new levels of ordering. This arises from the emergence of self-assembling proteins, like tubulin, providing microtubule linkage of cells. The new degree of order made available at such a stage is indicated by the coupling of chemical oscillators. When two batch reactors are physically linked in order to allow some chemical exchange to occur, the oscillators couple in a well-defined 'phase-locking' manner. The behavior of slime mould colonies also illustrates how new ordering features arise from such linkage of individual cellular entities.

A vast array of developments can now occur, with a variety of cell entities evolving in parallel to produce, at some stage, the fusion of different types. This process of endosymbiosis has been highlighted, in particular, by Margulis. Eukaryotes could have arisen in this way, via incorporation of mitochondria or chloroplasts into more primitive cells. Such fusion of differently ordered states provides yet another gateway for the further blossoming of higher-level cellular sophistication. The transition taking place is that from chemical evolution to biological.

CHEMICAL TO BIOLOGICAL EVOLUTION

The tight nexus between systems and the environment is maintained as evolution follows the axis of maximum complexity. Steady internal modification of chemical/physical character occurs, leading to the emergence of new features. Meanwhile, the environment evolves geologically and atmospherically. Interplay between the internal and the external determines the detail of development. Changes in the environment drive life-form change but the opposite also occurs. The availability of oxygen from the metabolism of plants opens the door for the arrival of mammals. At these stages, as emphasised by Dawkins, no element of the special accrues so far as particular detail of what has evolved is concerned. Moreover, specialised organs, like the eye may develop relatively quickly from the onset of essential cellular features, in this case sensitivity to light. But what emerges for any situation is a function of the nature of the particular time-line of the universe. Unique outcomes arise from the confluence of streams of developing events.

Dinosaurs ruled the earth for millennia, but became extinct, almost certainly because of a meteor impact 65 million years ago. Moreover, it was probable that this singular event provided the opportunity for the mammalian evolutionary branch to develop more dramatically, ultimately leading to humankind. Our emergence may be viewed as a function of a complex web of interrelated events. Whether or not we regard a meteor impact as an 'ordering' event is a moot point, but it should make us wary of definitions of order. Even chaos, as related to complex systems, contains order of certain kinds, since, for any system, components occupy a defined region of phase space.

In moving from chemical to biological evolution, the same ordering principles persist. The basic physical concept of openness (contingent upon 'enthalpy-entropy dissipation') extends to the biological domain by remaining an integral feature of component entities. Simultaneously it allows the environment to inform further evolution. Each environmental change sees variation in any species (via mutational dispersion), with one having the best overall survival features. Its total chemistry, in terms of 'openness', best matches the new environment, thereby ensuring its 'selection'. While in practice there is a complex trade-off of different features, integrated behavior remains correlated with degree of openness.

It does not appear necessary to be more sophisticated in identifying the physical basis of biological evolution. The details become increasingly complex, yet the driving forces remain the same. New order emerges to become genetically retained. Genes appear 'selfish' in lieu of chemical efficiency and lifetime ascendancy of phenotypic expression. Catastrophes like volcanic eruptions and asteroid impacts bring discontinuities and changes of direction. But the forces and flows of the universe continue.

Yet the emergence of the human mind ushers in a startling disruption to the natural balance of order and disorder established over aeons of time. Totally new features of matter ordering arise. Linked to manipulative capabilities, this mind fabricates an amazing array of spatially ordered objects, spawns patterned sounds in language and music and creates images of diverse meaning. In a dramatic turn-around, it becomes organiser of the environment that organised it. Even at a stage where it has already effected irreparable change by activating previously inert elements and totally new chemicals, an irresistible drive for greater control continues. Humankind enters the threshold of injecting into its environment its own genetically devised forms of life; forms that have not, *in toto*, emerged as contingent entities of the universe.

Given the insights offered by this study of ordering, such a pre-emptive venture may seem unwise. New characteristics so developed may endow the biosphere with unsuspectedly far-reaching consequences. The factors involved are far too complex to effectively analyse and monitor. A significant risk is therefore involved, especially in view of the profound underlying driving forces and tight nexus between development and environment identified here. The recent call for a 50-year moratorium on 'engineering' the human genome could well be expanded to cover release of any newly fabricated organism. This may be especially important for combinations of genetic material derived from widely divergent stages of evolutionary development.

Bibliography

Balasubramanian, D. and Rodley, G.A. [1991]. Incorporation of a Chemical Oscillator into a Liquid-Crystal System. *Journal of Physical Chemistry*. **95:** 5147–5149.

Conveney, P.V. [1988]. The Second Law of Thermodynamics: Entropy, Irreversibility and Dynamics. *Nature*. **333:** 409–15.

Dawkins, R. [1976]. *The Selfish Gene*. Oxford: Oxford University Press.
Dawkins, R. [1986]. *The Blind Watchmaker*. New York: Norton.
Eicke, H.-F. [1982]. Self-Organisation of Amphiphilic Molecules: Micelles and Microphases. *Chemia*. **36:** 241–246.
Eigen, M. and Schuster, P. [1979]. *The Hypercycle*. Berlin: Springer-Verlag.
Epstein, I.R. , Kustin, K. , De Kepper, P. and Orban, M. [1983]. Oscillating Chemical Reactions. *Scientific American*. March, 112–123.
Field, R.T. [1985]. Chemical Organisation in Time and Space. *American Scientist*. **73:** 142-149.
Gell-Mann, M. [1994]. *The Quark and the Jaguar*. London: Little, Brown and Co.
Gonda, I. and Rodley, G.A. [1990]. Oscillatory Behavior of a Belousov-Zhabotinskii Reverse Micelle System. *Journal of Physical Chemistry*. **94:** 1516–1519.
Hess, B. and Boiteux, A. [1980]. Oscillations in Biochemical Systems. *Berichte der Bunsen-Gesellschaft fuer Physikalische Chemie*. **84:** 346–351.
Lumry, R. and Rajender, S. [1970]. Enthalpy-Entropy Compensation Phenomena in Water Solutions of Proteins and Small Molecules: a Ubiquitous Property of Water. *Biopolymers*. **9:** 1125–1224.
Maddox, J. [1991]. Is Darwinism a Thermodynamic Necessity? *Nature*. **350:** p 653.
Mandelbrot, B. [1977]. *The Fractal Geometry of Nature*. New York: Freeman and Co.
Margulis, L. [1981]. *Symbiosis in Cell Evolution*. San Francisco: Freeman and Co.
Nicholas, G, and Prigogine, I. [1977]. *Self-Organisation in Non-Equilibrium Systems: From Dissipative Structures to Order through Fluctuations*. New York: Wiley.
Olmsted III, J. [1988]. in [Eds] Weber, B.H. , Depew, D.J. and Smith. J.D. , *Entropy, Information and Evolution*, Massachusetts: MIT Press Cambridge.
Orgel, L. [1992]. Molecular Replication. *Nature*. **358:** 203–209.
Prigogine, I. [1978]. Time, Structure and Fluctuations. *Science*. **201:** 777–785.
Rebek, Jr. , J. [1994]. Synthetic Self-Replicating Molecules. *Scientific American*. July, 34–40.
Schrodinger, E. [1944]. *What is Life?* Cambridge: Cambridge University Press.
Travis, J. [1994]. Hints of First Amino Acid Outside Solar System. *Science*. **264:** p 1668.

6. Molecular Biology – From *E.coli* to Elephant

G. Padmanaban* and M.S. Shaila**

Departments of Biochemistry and Microbiology & Cell Biology**,*
Indian Institute of Science, Bangalore 560 012, India.

The bacteria, *E.Coli* and the giant mammal, the elephant, cannot brook any comparison, other than sharing the first letter 'E'. But, our present state of knowledge clearly indicates that the two have several things in common, at the same time manifesting differences as large as their sizes would indicate. This unity in diversity concept is a result of the evolution of all life forms from a primordial cell. Krishna says in the Bhagawad Gita:

"Among cows I am the celestial cow Kamadhenu
Among Serpents I am Vasuki (King)
Among Nagas (special class of serpents) I am Ananta (Five headed)
Among quadrupeds I am the lion
Among birds I am Garuda (Kite)
Among fishes I am the alligator
I am the seed of all beings
I am the sustainer of all, having any face on all sides
or what will you gain by knowing all this in detail, Arjuna?
Suffice it to say that I stand holding this entire universe by a spark of any yogic power"

(Quote from different stanzas of chapter 10, Bhagawad Gita.)

The purpose of the quotation is not to propagate the Hindu philosophy on the origin of life, but to present an elegant way of recognizing a common thread linking all life forms. The possible origin of the primordial cell and how it could have given rise to the myriad of life forms constitutes an exiting story. It is an extraordinary story based on conjectures, assumptions and interpretations, because it is not easy, at this point of time to describe the course of events starting 3.5–4 billion years ago, when life probably originated on this planet. However, the molecular architecture and the functional attributes of the molecules in relation to the behavior of the present day organisms give a footprint of the past history that is clear enough to make reasonable predictions.

THE BASIC MOLECULAR PLAN

If life processes can be broken down to chemical reactions, then these reactions are brought about by proteins acting as enzymes to catalyze these reactions. This, along with structural materials in combination with other components provide a platform for the reactions to take place. Then, the basic plan becomes one of having a repository carrying the code to produce all the proteins. As a consequence, there has to be a machinery to decode this information so that the proteins are actually produced whenever needed. It is also imperative that the repository as well as the machinery for decoding are transmitted from parent to progeny. It is worthwhile to remember that the repository has all the information including the code for the machineries involved in decoding and hereditary transmission of information. Francis Crick enunciated the central dogma in molecular biology, which in its updated form can be written as:

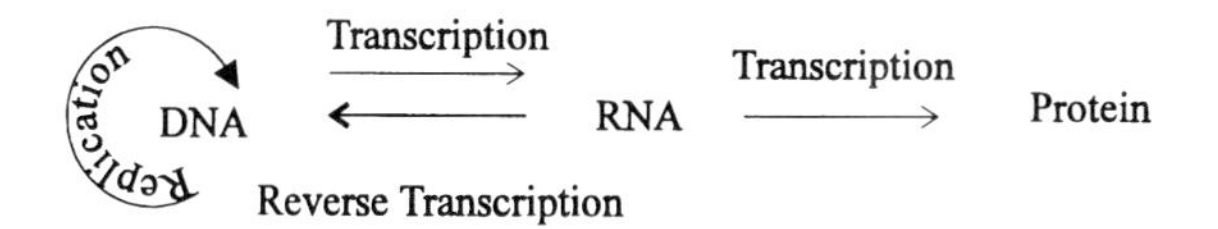

The process of transcription has become a key step in various stages of growth, development and differentiation. This is the process by which the information, present in DNA as genes, is decoded to give rise to RNA, which is then translated to give the protein. It is of importance to remember that all genes are not transcribed at the same time. There are house-keeping genes, whose products are required at all times of the life cycle of an organism, and there are also genes which come into play at crucial stages of development and differentiation of a cell and may even be active only for a short period. The gene product may send in a cue or a signal that will trigger a series of reactions and once its function is achieved, the gene may become silent. Thus, gene activation and repression are processes brought about in response to environmental stimuli mediated through the cell. The process of signal transduction by which external stimuli are transmitted to the DNA, whereby specific genes are activated or shut-off is an exciting field of investigation. The stimuli can range from physical parameters such as light and heat, to chemical parameters such as vitamins and hormones, and to physio-pathological states such as anxiety and infection. All these signals are eventually translated into molecular signals influencing gene transcription. Although, regulation of gene transcription can occur at different levels, one important facet is the role of transcription factors. These are protein in nature, function as controlling elements and are specifically able to interact with DNA sequences. These DNA controlling elements flank gene sequences and their interaction with transcription factors plays an important role in gene activation or suppression. The various signal transduction mechanisms often influence the levels or active states of the transcription factors.

The information carried in the messenger RNA is translated into protein on the basis that each amino acid has a specific triplet nucleotide code. Interestingly, each

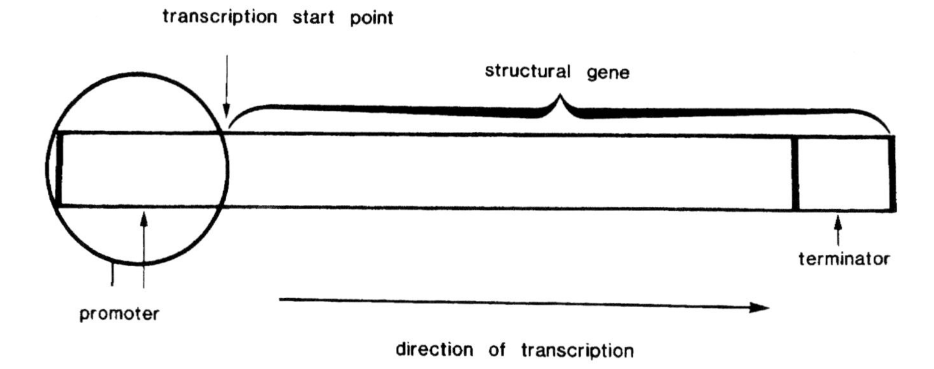

Figure 1. The basic prokaryotic transcriptional unit

triplet codon in the messenger RNA is recognized by an anticodon present in another RNA molecule referred to as transfer RNA, the latter carrying a particular amino acid. The successive reading of the different triplet codons in a non-overlapping fashion ultimately brings the different amino acids into proximity, the first amino acid being linked to the second through a peptide bond. The synthesis of a protein needs an elaborate machinery of ribosomes and a number of protein factors catalyzing the various steps. The proteins thus made can undergo further modifications and are distributed to different parts of the living cell to carry out a variety of functions.

THE PRIMORDIAL CELL

In today's world, chemical evolution has perfected the nature of molecules serving as information repositories and those involved in signal transduction, information decoding and propagation. Concomitantly, biological evolution has been perfecting life forms to acquire attributes for a purposeful and intelligent life. Several questions arise; How did the primordial cell evolve? How did multicellular organisms evolve? What is the common thread linking this evolution? Can life be defined in cellular or molecular terms? There has been a knowledge explosion in many of these areas, although there are still formidable gaps to be filled in.

It is held that the prebiotic condition on earth was very unstable with violent electric discharges, ultraviolet light and very little free oxygen. Laboratory experiments indicate that these conditions could have given rise to compounds such as formaldehyde (HCHO) and hydrocyanic acid (HCN), which can undergo further reactions in an aqueous environment to give rise to amino acids, sugars, purines and pyrimidines. These are the basic building blocks for the biopolymers, proteins, carbohydrates and nucleic acids. High concentrations of each of these components can give rise to the appropriate polymer, the motivating principle being the ability to generate more of the same kind. This autocatalytic potential would have favoured the formation of nucleic acids rather than proteins. This inference is based

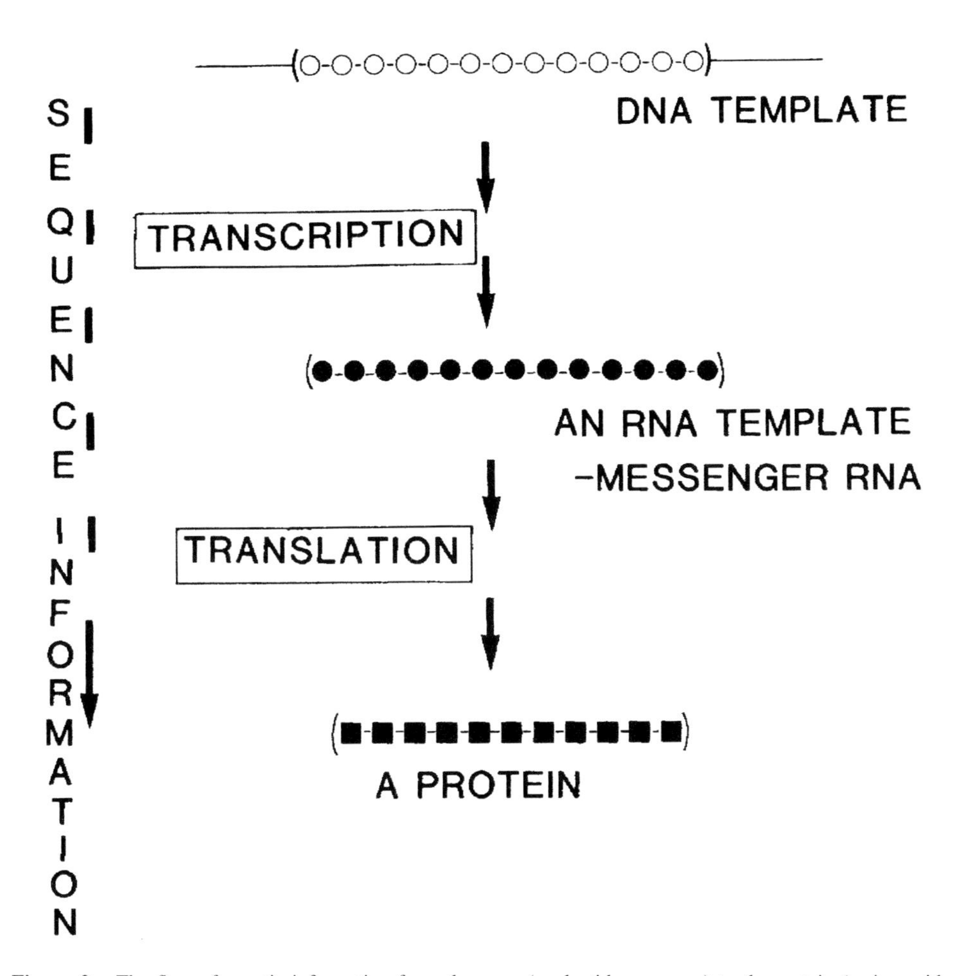

Figure 2. The flow of genetic information from the gene (nucleotide sequence) to the protein (amino acid sequence)

on the structures of these polymers. Nucleic acids consist of purines (A and G) and pyrimidines (T and C), with A capable of pairing with T and G capable of pairing with C. These bases are attached to a sugar-phosphate backbone and in view of the base pairing capabilities, a single strand containing several bases (purines and pyrimidines) can give rise to a complementary strand. Thus, there is a selective advantage for the duplication of the primitive nucleic acid on its own template. Although, proteins which are composed of 20 amino acids are superior catalysts compared to nucleic acids which contain only 4 bases, the former lack the mechanism to reproduce. Thus, the early polymers would have been mostly nucleic acids with some polypeptides formed by the fusion of amino acids.

Evidence suggests that the primordial informational macromolecule could have been ribonucleic acid (RNA) rather than deoxyribonucleic acid (DNA). RNA is single-stranded, unlike DNA which is mostly double stranded and the RNA molecule in the present day environment manifests certain properties indicative of its role as the early informational macromolecule. RNA besides being capable of fully reproducing itself using templating mechanisms, is also capable of folding on itself using the same mechanism to give a partial double stranded character. These flexibilities enable the RNA molecule to assume a catalytic function, although not as good as that of proteins. The RNA molecule does have the property of cutting and joining (splicing) other RNA molecules, including self-splicing. These properties would have generated different RNA molecules, where the base sequences are different, and therefore endowed with diverse information content. The order of the base sequence would decide the information content. Ultimately, the demands of survival and evolution would have put a severe stress on the RNA molecule to change as the single-stranded nature of the RNA molecule leaves it susceptible to attack by other ions leading to easy degradation. Besides, the presence of hydroxyl groups on the 2'C of the sugar in the sugar-phosphate backbone renders it susceptible to hydrolysis. Thus, DNA with its double-stranded structure should have evolved by reverse transcription (RNA→DNA). This double-stranded nature with each strand being complementary to the other, stabilizes the DNA. Besides, the hydroxyl on the 2'C group of the sugar is replaced by hydrogen and this renders the sugar phosphate backbone resistant to hydrolysis. The stability of DNA is at the expense of the versatility of RNA. The information content in DNA is well protected and needs to be read, transcribed and translated, since DNA by itself cannot execute the functions. Thus, the pathway of information flow, DNA→RNA→ protein, by mechanisms described earlier must have evolved. The double stranded DNA, the two sister strands being complementary, also presents an elegant strategy for the segregation of each of the two strands to daughter cells, with each strand acting as a template to form the new complementary strand. Thus, transmission of information from parent to progeny is in-built in the structure of DNA. RNA acts as a true intermediate in the transfer of information from DNA to protein. Todays RNA is not simply a relic of the primordial RNA, but has several vital functions to perform, mostly in the realm of transfer and generation of information from DNA. RNA splicing is capable of generating newer sequences producing different sets of information from a single sequence of DNA. Translation of RNA into protein, by the steps mentioned earlier, follows certain rules. The RNA molecule has to be read in one particular orientation (5'→) and the triplet codon read from a fixed initiation point. The non-overlapping mode of reading the triplet can still give rise to three reading frames in one direction and only one of the reading frames provides sense information, thus giving rise to a protein product with the correct amino acid sequence to confer a particular catalytic and/or structural property expected of that protein. The elaborate translation machinery ensures these aspects.

It is obvious that effective molecular interactions can take place only when adequate concentrations and proximity of the interacting molecules are ensured. Thus, interactions between nucleic acids and proteins are promoted in a closed environment by resorting to compartmentalization. Amphipathic molecules can form a limiting membrane and isolate the reacting components from the environment. The primordial cell could have formed by the spontaneous assembly of phospholipid molecules in the prebiotic soup to enclose the RNA, some protein and other molecules. The membrane has to function not only to isolate the components inside, but also as a channel for communication with the outside.

EVOLUTION OF MULTICELLULARITY

Geological evolution has necessitated the evolution of multicellular organisms not only for survival under a wide variety of conditions, but also to evolve intelligent means of survival and an improvement in the quality of life. However, unicellular organisms have their own niche and survive under a wide range of conditions existing on earth, constituting a significant proportion of the biomass. Unicellular organisms often cannot tolerate drastic changes in the environment for which they have developed a machinery for survival. Within limits, these organisms do manifest a regulatory potential through molecular mechanisms of induction and repression. Thus, a change in the nitrogen or carbon source can lead to expression of new proteins to metabolize the new carbon or nitrogen source. Unicellular organisms can also shut off the expression of a set of genes required to produce a metabolite, if such a metabolite is available as a nutrient to the organisms. Extremophiles can survive under conditions of high and low temperatures. Unicellular cells may also associate with each other to form a clone of cells and this cellular cooperation ensures better survival. For example, under conditions of starvation, fruiting bodies are formed within a clone which are present in spores that can lie dormant. Under appropriate conditions the spores can germinate and give rise to vegetative growth. Despite all these strategies, unicellular organisms can only survive in a given environment and multicellularity bestows the ability to diversify the potential of a given organism for different functions and program for life at high levels of cellular co-operation and homeostasis. Multicellularity involves cooperation among different types of cells, some of which become specialised to perform a particular function. The root cells in plants are involved in uptake of nutrients, whereas the leaf cells carry out photosynthesis. The red cells in animals carry oxygen, whereas the myocytes are involved in muscle contraction. Thus, different types of cells specialise to carry out specific functions and are referred to as terminally differentiated cells. In between, are the house-keeping cells which have to perform a variety of mundane functions such as metabolism to utilize and store energy and generate a variety of metabolites.

In addition, each cell of a multicellular organism has evolved functions that are compartmentalised, perhaps as acquisitions from the prokaryotic cell. For example, mitochondria, the powerhouse of a cell, have evolved in the eukaryotic cell through endosymbiosis with a bacterium. Similarly, it is held that plant cells could have occluded cyanobacteria with photosynthetic capabilities, to have given rise to chloroplasts. While the similarities between these organelles and the prokaryotes is striking at the molecular level, there are also differences indicative of the changes brought about in these acquisitions in the process of their integration

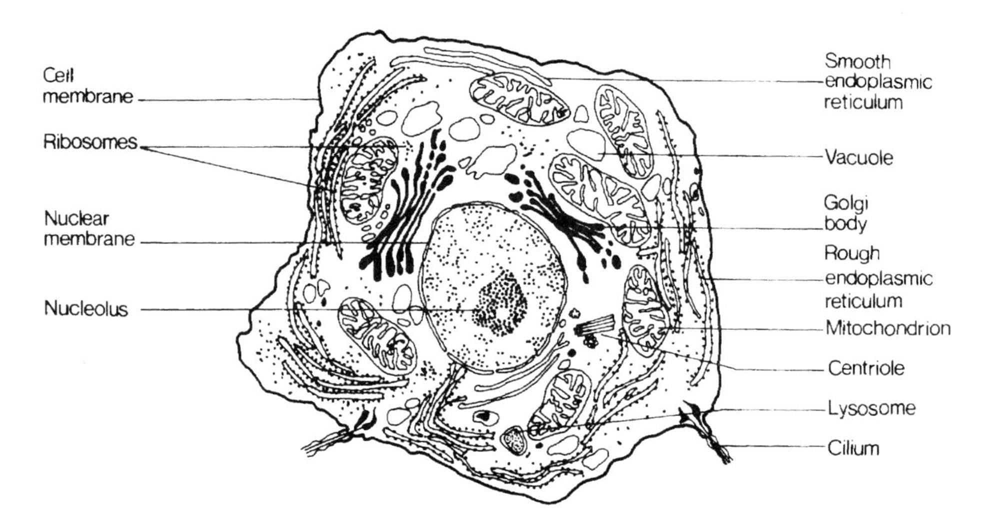

Figure 3. Schematic representation of a typical eukaryotic cell.

with the eukaryotic cell. An important feature of the eukaryotic cell is the compartmentalization of the genetic material in the nucleus, which is bounded by a double membrane. There are also other membrane structures such as peroxisomes, endoplasmic reticulum and cytoskeletal elements. Since the nucleus is the seat of genetic material, it is the site for transcription (DNA→RNA), while the endoplasmic reticulum studded with ribosomes is the site for protein synthesis. The messenger RNA made in the nucleus is therefore transported to the endoplasmic reticulum where the translation machinery is located. The mitochondrion synthesises ATP by the oxidation of food metabolites and this stored energy is utilized by all cells for a variety of functions. In the multicellular eukaryotic cell, there is coordination of function within and between cells. Thus, the complexity of higher organisms at the cellular level is two-fold. The evolution of eukaryote from a prokaryote involves segregation of functions by compartmentalization within a cell. Secondly, the evolution of different types of terminally differentiated cells with specialized functions, calls for intercellular cooperation. This is brought about by physical cell-cell interaction as well as through long distance communication, mediated by appropriate signalling mechanisms transmitted through chemical messengers.

C VALUE PARADOX

The acquisition of diverse functional capacities of multicellular organisms calls for the involvement of many new proteins and therefore, the evolution of many new genes. Thus, there is the need for an increase in DNA content. The total amount of DNA in an organism estimated for the haploid genome (eukaryotes are diploids with two copies of each chromosome) is referred to as its C value. The C value of prokaryotes and eukaryotes shows a wide range from 10^6 to 10^{11} base pairs. While the increase in functional capabilities does require more genetic potential, the increase in the latter is not proportional to the increase in the number of genes. For example, within amphibians, the DNA content ranges from 10^9 to 10^{11} base pairs and it is difficult to imagine that there could be 100-fold difference in the number of genes in this genera. It turns out that a significant proportion of DNA in higher eukaryotes has no coding function. These non-coding sequences occur in the form of intervening sequences between and within coding regions. A substantial proportion of the non-coding sequences outside the coding region are repetitive in nature. One reason suggested for the origin of such DNA, is the drive for

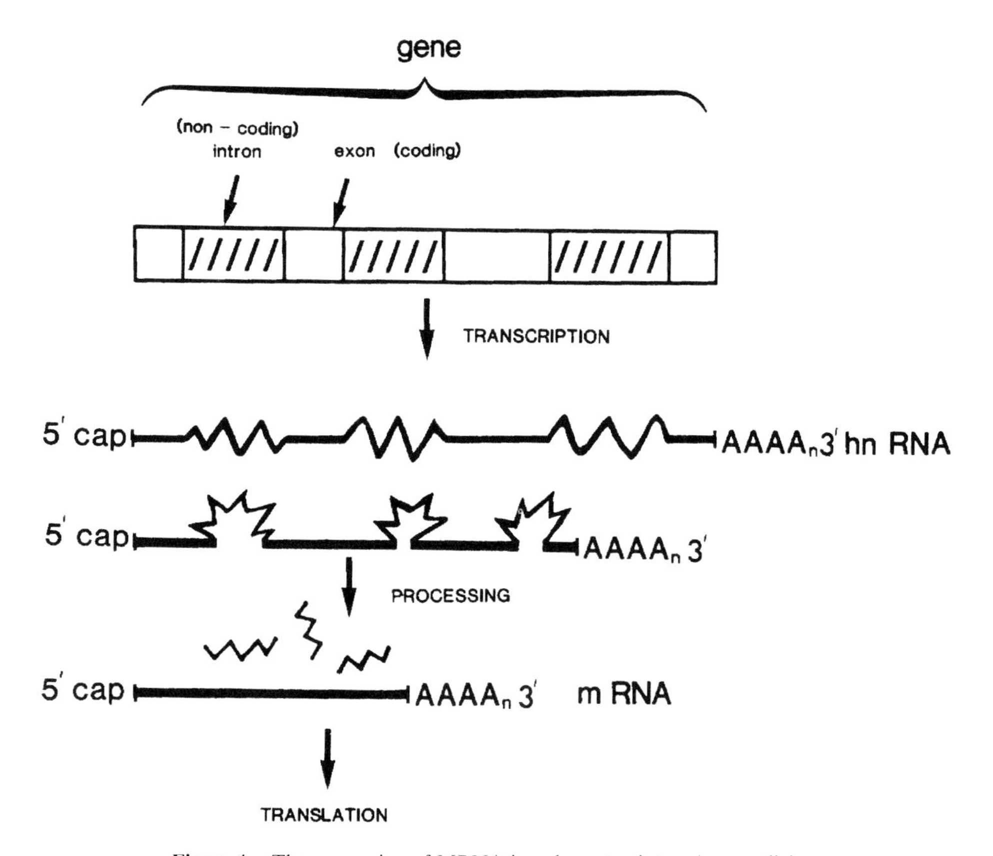

Figure 4. The processing of MRNA in eukaryotes-intron/exon splicing.

reproduction of sequences of the same kind, as suggested earlier for the evolution of polymers from monomers.

The functional role of large chunks of non-coding DNA is not well understood. It is difficult to imagine that large chunks of DNA are just present without a function and it has become clear that non-coding sequences flanking the coding sequences (genes) play a role in regulating the expression of the genes. Even sequences a large distance from a gene seem to have a remote control and often such sequences come into proximity of the gene due to bending or looping out of the intervening sequence. While the functional role of large tracks of repetitive sequences is not understood, recent studies indicate that in some of the genetic disorders the extent of repeats changes dramatically indicating that they must be playing a role in normal homeostasis. Non-coding sequences within genes, referred to as introns, are spliced out at the RNA level to give the correct messenger RNA that has continuous coding information. Introns provide the hot spots for change, while the coding sequences (exons) are generally protected. Thus, the generation of the old protein or its functional activity is ensured, while evolutionary experiments are conducted to generate newer proteins. Since introns determine the course of splicing, mutated introns can alter the splice site and thus generate newer splice sites and thus generate new reading frames. Despite all this information, the C value paradox is an enigma. This stems from the fact that the genome of higher eukaryotes appears to use only a small proportion of its DNA for coding function. Is it possible that the non-coding DNA actually constitutes the alphabet for a language, which is obviously quite different from that of the coding region? If so, what is this language and what does it convey? Is it likely that nature would have wasted so much of its energy and resource to make selfish DNA without a purpose? These questions still need to be answered.

DROSOPHILA – WORK HORSE OF DEVELOPMENTAL BIOLOGY

The development of an organism from the embryo has close parallels to its evolutionary history. As already discussed, the genome has undergone changes with evolution, all of which are not explainable. With the issue of Development, the question is how different tissues and organs arise from the same genetic information present in the embryo. The fruit fly, Drosophila, has turned out to be an excellent model to study the molecular basis of development. A brief description of the morphological changes during development would be as follows : The fertilized egg cleaves to form many smaller cells referred to as blastomeres. These become organized and in the outermost regions of the embryo, the blastomeres form a seal through tight junctions, isolating the interior of the embryo from the external medium. Osmotic changes then lead to development of fissures inside the embryo and these enlarge to form a single cavity. The embryo is now referred to as the blastula. The cells on the exterior of the blastula organized on a sheet of epithelium start undergoing complex movements. An invagination process takes place in which cells in the exterior move inside to form the embryonic connective tissue referred to as mesenchyme. This process is known as gastrulation. The mesenchyme gives rise to a three-layered structure, the ectoderm, endoderm and the mesoderm. These three layers respectively give rise to the epidermis on the outside, gut on the inside with connective tissue and muscle between them.

Segment formation in Drosophila has been the subject of intense investigation as a model to understand the molecular basis of morphogenesis, a process by which new organs are formed from a homogeneous looking embryo. The fly consists of a head, three thoracic segments and eight or nine abdominal segments. With gastrulation (embryo at 5–8 hours), segmentation is clearly visible. The segments look alike, but the body plan is already laid and the fate of each segment is defined at the molecular level. In fact, there is asymmetry in the seemingly homogeneous looking embryo and even in the egg cell derived from the mother. In the egg cell, yolk is concentrated in the lower end called the vegetal pole. The upper end is called the animal pole.The drosophila forms a syncytial embryo, in which many nuclei are present within a large cytoplasm, providing scope for easy diffusion of components and establishment of gradients. Two maternally coded proteins, Bicos and Nanos, diffuse through the syncytial embryo from the posterior and anterior ends respectively. Therefore, a protein gradient is set up that is responsible for the activation of

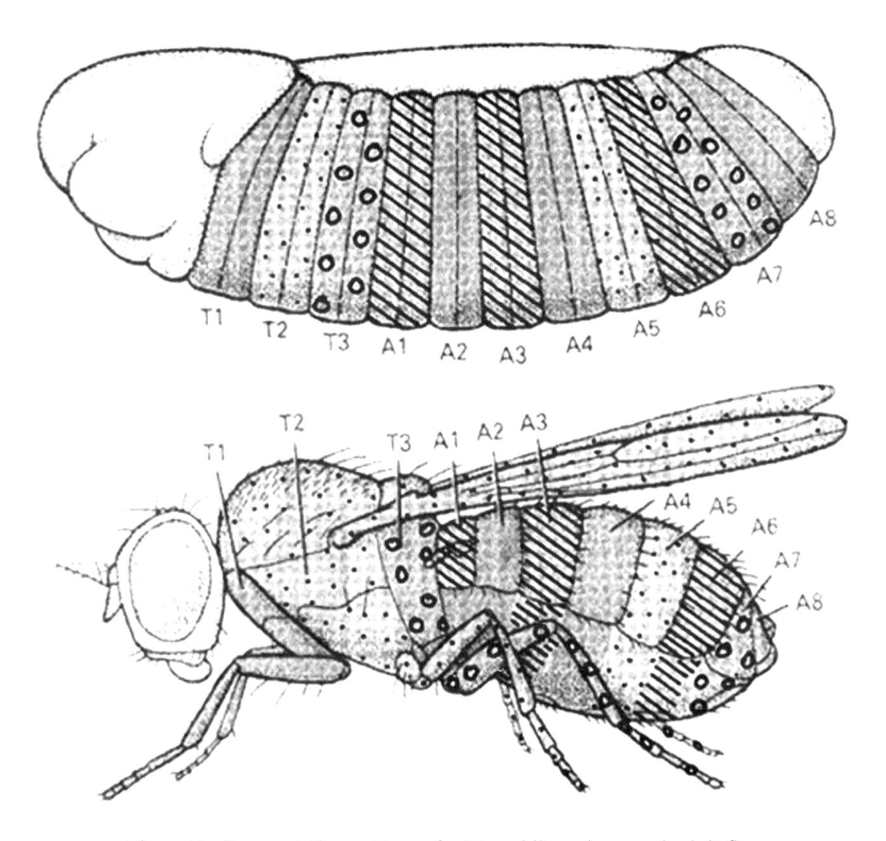

Figure 5. Segmentation pattern of a Drosphila embryo and adult fly.

a hierarchy of genes in a position dependent manner. In the gene hierarchy, gap genes determining the spacing between embryonic segments are the first to respond. The gap gene products are essentially transcription factors that regulate the expression of pair-rule genes, which decide the periodic patterning of the stripes.The pair-rule gene products are exclusively transcription factors, which in turn regulate the expression of segment polarity genes. These determine patterning within each segmental unit and help to refine the borders established by pair-rule genes. The segment polarity gene products are transcription factors and some code for proteins involved in intracellular communication and signal transduction. Concurrent with segmentation, the expression of homeotic genes, the highest in the hierarchy, gives unique identities to each segment. Thus, the Drosophila embryo undergoes progressive subdivision and identities are assigned to each segment by the homeotic genes. The homogeneous looking Drosophila blastoderm contains an accurate representation of the larval body plan.

CONSERVATION OF DEVELOPMENTAL STRATEGIES IN EVOLUTION

Biodiversity is a gift of nature, which has chosen to conserve the developmental plans to generate the diversity. Nature has not resorted to introducing totally new machineries but redeploys already known machineries in different contexts. The knowledge acquired by the primordial organism to make head and tail is conserved, but elaborated to introduce additional complexities in evolution. For example, the principle of asymmetry between cells in the embryo, responsible for establishing different cell lineages and fates, is maintained but through different mechanisms. As already mentioned, even the egg cell is asymmetric in drosophila. In amphibians, soon after fertilization, there is rotation of the cortex of the egg relative to its core, adding to the asymmetry. Mammalian eggs are essentially symmetrical unlike amphibians. Most animal embryos become cellularized before asymmetry is seen to be established, whereas in drosophila the syncytial embryo permits free diffusion of maternal proteins establishing a gradient of morphogens to activate specific genes in a position specific manner. In animal cells, positional, temporal or lineal (segregation of regulatory information asymmetrically during cell division) strategies can be followed to regulate cell fate decisions. Signal transduction across cell membranes by morphogens in the cellular environment or due to cell-cell interaction can trigger of gene cascades. It is known, for example, that mouse embryo cells removed from the normal influence of their neighbours develop into terato-carcinomas but revert back to normal development when implanted back into a normal embryo. The morphogens as such have not been identified but substances such as retinoic acid are the likely candidates.

How is cell-cell contact established? One obvious link is through the expression of cell adhesion molecules. For example, in Drosophila, a gene referred to as wingless (wg) activates another gene referred to as Armadillo through two other genes. Armadillo actually codes for cell adhesion proteins. An interesting example of signal transduction is the expression of the 'Ras' pathway in Drosophila. Ras a proto-oncogene, has a counterpart in the gene referred to as sevenless. This gene is needed for the development of a neuron in the fruit fly. The gene codes for a tyrosine kinase receptor, which mediates a signal to phosphorylate proteins on specific tyrosine residues and is referred to as the Notch pathway. The reception of the signal, however, depends on the previous history of the cell. In this particular case, the previous history pertains to the distribution of a protein called Nuans. There is asymmetric distribution of this protein, with some cells having more and others less of this protein. In cells having a greater amount of Nuans protein, the Notch pathway is inhibited and these become neurons. In cells having inadequate amounts of Nuans, the Notch pathway is not inhibited and these do not become neurons but become accessory cells. Thus, asymmetry between cells plays a role right through from embryogenesis to morphogenesis.

It is also interesting to note that the same families of molecules that guide the earliest stages of embryogenesis also play a role later in morphogenesis. These are conserved in evolution and play similar developmental roles from fly to fish and rodents to race horses. An interesting example is the family of genes related to wingless (wg) in Drosophila mentioned earlier. Expression of this gene actually helps in the organization of the 14 segments in the embryo that ultimately form the major components of the adult fruit fly's body. Later on, the gene also plays a role in wing development, since mutation in this gene leads to defective wing development. The vertebrate relatives of the wg gene are known as the Wnt genes. Wnt-3a is needed for mesoderm formation during gastrulation. Wnt-1 is implicated in vertebrate brain development. Wnt-4 is needed for kidney formation. Yet another example is the 'ras' pathway mentioned earlier. The protein product of this ras oncogene, guides eye development in Drosophila and vulval development in the nematode *C.elegans*. That there can be a connection between oncogenesis, eye development in the fly and vulva formation in the nematode is remarkable. Thus, redeployment of the same or similar genes in different contexts within the life span of an organism or in an evolutionary context brings about remarkably different end results, which can range from the development of a new organ to the growth of cancerous tissue. It is a mystery of nature as to how it regulates the expression of similar genes to orchestrate diversity. There is no denying the fact that an enormous amount of information on the gene component and its regulation of expression in different contexts is available, but the mystery has only deepened! Opening up of one tunnel leads to another tunnel and to see light at the end of the tunnel is like catching up with the rainbow at the horizon.

HOMEOTIC GENES

These genes functioning at the highest level of hierarchy in gene cascades leading to organ formation behave as molecular address labels for the cells in each parasegment in Drosophila. If the address labels are changed, the cells in one parasegment behave as though they are located elsewhere. In Drosophila, the homeotic selector genes lie in two tight gene complexes referred to as Bithorax complex and Antennapedia complex. The former controls the identity of abdominal and thoracic segments of the body, whereas the latter controls that of thoracic and head segments. Thus, mutation in Bithorax would lead to an additional pair of wings in the place of halteres, while mutation in Antennapedia could lead to the sprouting of legs in place of antenna.

The homeotic selector genes code for highly homologous gene regulatory proteins containing a conserved homeobox sequence. This sequence consists of a DNA-binding homeodomain of 60 amino acids. As already indicated, transcription factors coded by egg polarity and segmentation genes as well as others in the hierarchy, influence the HOM complex expression which acts as an interpreter of the positional information conveyed by the various factors. The cell has a memory for the positional information, recorded by the homeotic selector genes resulting in the coordination of cell signals to generate specific patterns. For example, the cells from two pairs of limb buds in the embryo eventually giving rise to leg and wing respectively, appear very similar. If a piece of the undifferentiated tissue from the base of the leg bud region, which normally gives rise to the thigh region, is grafted into the tip of a wing bud, it develops into a toe. Thus, first of all the leg bud is already committed to become part of a leg and therefore, cannot become part of a wing even though it is grafted into the wing bud, since its positional value is different. However, the different wing bud cell environment influences the leg bud cells to form the tip (toe) rather than the base (thigh).

Homeotic selector genes first get activated in the blastoderm stage but play a continuous role to decide the course of morphogenesis. In view of their conservation in evolution, the head to rear axis of vertebrates is homologous to that of the anterio-posterior axis of the fly. A remarkable feature of these genes is that they are ordered on the chromosome in exactly the same order in which they are expressed along the axis of the body. This is true of flies and vertebrates. Not only the gene order is conserved, but also the regulatory enhancer sequences of these genes are ordered and conserved. All these argue for positional regulation of cells, responding to a gradient of morphogens, set up from one end to another. However, as already indicated, organisms do follow positional, temporal and lineal strategies to set up a conserved pattern of *Hox* gene expression. But, despite the remarkable diversity of embryos and the differences in the modes of initiating development, as the embryos establish their body plans and initiate morphogenesis, *Hox* gene expression is uniformly triggered.

Hox genes have probably evolved from a common precursor and undergone further modification in evolution. As already indicated Drosophila has two complexes; Bithorax and Antennapedia housing eight genes. Vertebrates have four complexes, each having nine to ten genes. Arthropods and nematodes have only one complex. Duplication of *Hox* genes perhaps occurred after the divergence of vertebrates and arthropods. The *Hox* genes must have replicated several times in the evolution of the vertebrate line. This has given the extraordinary ability to vertebrates to evolve new body patterns.

THE HEALING TOUCH

All this progress is exciting. Although, several facets of Developmental Biology are not understood, the basic plan of nature to develop and evolve body structures in organisms is in perspective, although molecular details are complex. It is also amazing that the basic plan is conserved in evolution. That is why the study of simpler genomes has given a wealth of information on complex genomes. In fact, this unity concept has evoked the rather overstretched observation, "What is true of *E.Coli* should be true of Elephant." Despite its limitations, this is the basis for using prokaryotic and eukaryotic models to understand not only basic processes of development and differentiation, but also of disease and death. This has paid rich dividends leading to the discovery of drugs and pharmaceuticals to protect against infection. This knowledge has formed the bedrock of the whole area of recombinant DNA technology that has revolutionalized biotechnology.

Systemic and genetic disorders pose formidable challenges and one ultimate end result of many of these disorders stems from the inability of adult cells to divide. If only adult cells could have the capacity to regenerate like the embryonic cells, disorders as diverse as diabetes, blindness, heart failure, Alzheimer and Parkinson diseases and many other neurodegenerative disorders, healing of wounds and broken bones will have a hope for cure. This is the basis on which dramatic cures are expected by transplanting embryonic cells on worn out tissues. This, of course, poses technical problems and raises ethical issues as well. The mass of information on all the various protein factors regulating morphogenesis has given rise to the hope of finding an alternate approach to embryonic cell transplantation. This rests in providing the necessary growth factors to adult cells to enable them to divide. Thus, a range of factors from erythropoietin to Nerve Growth Factor are being examined towards repair of damaged tissues. How to provide these factors *in vivo*?

Ultimately, the approach has to be in terms of gene therapy, where the appropriate gene has to be targeted to the recipient cell and let the gene express *in situ*. Gene therapy of the zygote or the embryo will pose at least as many ethical problems as embryonic cell transplantation. Human embryo research is a very sensitive issue and gene manipulation of such embryos is beset with social, ethical and technical concerns. However, somatic cell gene therapy, in which tissues such as bone marrow, liver, skin, lung, etc. can become targets for gene therapy will become as much an option as kidney transplantation or heart valve replacement. The stem cell of the bone marrow is a sought after target for gene therapy of several blood borne disorders, since it can repopulate other cells derived from the progenitor cell. However, the paucity of an adequate number of stem cells and the efficiency of the targeting process, which often involves the use of retroviral vectors, are still major technical hurdles to cross. More recently, direct injection of genes into tissues, muscle and skin cells in particular, have been shown to elicit the production of the desired proteins. While, the efficiency of the process still needs to be augmented, there is a lot of excitement in this approach to use DNA or the gene as a pharmaceutical. Further, injection of genes for heterologous proteins has been shown to elicit antibody formation and provide cell-mediated immunity. Therefore, a whole new field of 'DNA vaccines' has caught the imagination of many research groups.

While many systemic disorders stem from the inability of adult cells to regenerate, cancer poses the unique problem of uncontrolled cell division and growth. All the molecular studies on development have clearly identified cancer as a problem of aberrant development. The very same growth factors, necessary for normal growth and development, with changes brought about by mutation, translocation to a different locus etc. cause the uncontrolled growth witnessed in cancer cells. It is amazing that so many regulatory circuits are all in place to facilitate growth and development and that they function by and large in a flawless fashion, since primary molecular differences responsible for normal and abnormal growth do not appear to be very large. Man's destructive intervention with the environment has often led to a breakdown of this sensitive balance, an aberration and insult at the cellular level far exceeding the potential for homeostatic self-regulation. Since a cancer cell is a close relative of the normal cell, the problem has always been to selectively kill the former. All approaches to kill cancer cells would also affect the surrounding cells. Once again, gene therapy holds promise in this area as well, since attempts are underway to target genes specifically to cancer cells. One approach is to use tumor infiltrating lymphocytes (TIL), which normally home in on cancer cells to destroy them. If a toxin gene can be targeted to the cancer tissue through the TILS, then the expressed toxin can selectively kill the cancer cells without affecting normal cells.

There is an explosive increase in our understanding of molecular defects in genetic disorders. This is because of the development of recombinant DNA technologies to clone and analyse large chunks of the human chromosome, subsequently to carry out fine mapping to localise the defect to a narrow region and finally demonstrate the changes at the DNA sequence level. The human genome project which aims at sequencing the entire genome of 3 billion base pairs will lead to the identification of the exact loci for all genetic disorders, which are estimated to number over 3000. In this context, the study of the smaller genomes of small organisms becomes relevant in view of the conservation of several developmental strategies.

LIFE IS MORE THAN ALL THE MOLECULES PUT TOGETHER

The reductionist approach of molecular biology has indeed broken down life processes into myriads of chemical reactions. This has helped in understanding several facets of growth and development and the knowledge has been very useful in evolving strategies to combat disease, promote health, grow more food and regulate population. Genes are being discovered that can account for higher functions such as intelligence, memory, emotions and behavior. The awesome power of the brain is being interpreted in terms of neural networks. Firing of neurons and release of neurotransmitters visualized to underlie many perceptions of the brain. It would, however, be difficult to interpret the self will of an individual to perform an act or a decision to change course in a split second in molecular terms that is the cause rather than the result.

Molecular Biology is at its pinnacle, deciphering genome after genome from *E.coli* to the Elephant. Where is it leading to? It is true that homologous genes are being discovered that have been conserved in evolution. It is true that the basic developmental plan conserved in evolution has been understood. No one can deny that such knowledge has been useful in deciphering strategies for improving the quality of life. But, one is not exactly comfortable that all this progress is enriching human endeavor to equitably benefit all mankind. There is a nagging feeling that man is trying to play God and introduce new facets of imperialism. Would populations be classified as superior and inferior based on genome maps? Would employers demand to see gene maps to know about fitness of individuals? Economic dominance has led to intellectual dominance, since rich countries are able to attract the best brains and provide adequate infrastructure to spearhead gene wars and win all the patents with potential biomedical and agricultural applications. The poor nations may be rich in biodiversity and gene pools, but their genes are being exploited elsewhere. If this world is to be a reasonable place for all to live with dignity, it would be necessary to share knowledge and create wealth not only for the rich but also for the poor. A new world order based on sharing of mutual concerns, understanding of each others problems and making economic progress with a human face has to evolve.

7. Immunology: The Immune System and Beyond

Graham R. Flannery

School of Genetics and Human Variation,
Faculty of Science and Technology, La Trobe University.

INTRODUCTION

An Historical Perspective

A little over one hundred years ago, the Russian zoologist Elie Metchnikoff, enjoying a self-imposed early retirement (he was not yet forty), observed the fate of rose thorns introduced into starfish larvae. He saw scavenger cells (phagocytes) engulf and chemically digest the foreign particles, rendering them harmless. The observations rekindled his interest in research and resulted in his sharing of the Nobel Prize for Medicine 26 years later. They also set in train a major controversy concerning the body's defence against infectious disease.

The early history of *immunology* (the study of the immune system and of the immune response) follows essentially the history of our understanding of *immunity* to infection and of our attempts to induce it. The concept of immunity is a very old one: the writings of Thucydides describe the resistance to further infection of plague survivors and their role as carers for the newly infected in ancient Greece. Attempts to induce immunity also have a long history: efforts at protection against smallpox (*variola*) were described over eight hundred years ago and included the Chinese practice of inhalation of powdered scabs (as prophylaxis against facial scarring in marriageable young women) and the Turkish practice of introducing pus into wounds in the skin. The latter (empirical and often fatal) 'variolation', reached England from Turkey in the eighteenth century. That a similar, but benign disease, cowpox (*vaccinia*), might protect against the dreaded smallpox was a widely held belief in rural England, substantiated by Jenner in 1798 (in an experiment guaranteed to have him struck from the Medical Register today). In spite of the potential now for a safe preventative against one of the world's greatest scourges, *'vaccination'* was only reluctantly accepted by the prevailing medical establishment. Immunisation (and immunology) proceeded slowly.

Almost a century after Jenner's vaccination, Pasteur's work led to the discovery of the *in vitro* attenuation of infective agents and the rendering safe of immunising organisms by treatment such as heat and chemical fixation. [Pasteur's development of an anthrax vaccine, first demonstrated in 1881, is regarded by many as the beginning of the discipline of immunology.] A major advance, however, was the passive transfer of pre-formed immunity from already immunised individuals via blood serum, discovered by von Behring and Kitasato (and for which they received the Nobel Prize in 1902). The 'antitoxins' (*antibodies*) responsible for such antibacterial immunity were then shown by Ehrlich, Bordet and others to be inducible against foreign proteins additional to those of microbes. With the clarification of the interdependence of antibodies and the complement proteins (described later) for which Bordet was awarded the 1919 Nobel Prize, the essentials of (antibody-dependent) *humoral immunity* had been established.

Thus it was that Metchnikoff's arguments for a cellular (*phagocyte*) basis to disease resistance – *cell-mediated immunity* – flew in the face of the established wisdom: immunity had been demonstrated to be a function of serum factors (humors) and not of cells. [Ironically, it was the great proponent of humoral immunity, Ehrlich, who shared the Nobel Prize in 1908 after many years of violent disagreement.] The discovery, however, of the opsonising properties of antibodies (the enhanced phagocytosis of antibody-bound material) signalled the beginnings of the marriage between humoral and cellular immunity. Indeed Almroth Wright argued that the function of antibodies was simply to facilitate the action of phagocytes. The uncertainty of the relationship (but also its pervasiveness in clinical medicine) is exemplified by Shaw's fictitious Sir Colenso Rigeon (speaking for Wright) in *The Doctor's Dilemma* (1908): '. . . . . *a suitable antitoxin. . . . . The phagocytes are stimulated; they devour the disease; . . .* ' Appreciation of the real, in fact central, role of phagocytes in immunity, the complex inter-relatedness of the elements of humoral and cellular responses, and the repertoire of yet to be discovered cellular and non-cellular components was, however, fifty years away. Nevertheless, the next two decades witnessed the firm establishment of immunization procedures designed to elicit specific therapeutic immune responses, the impetus to such work slowing only temporarily with the advent of the sulphonamides, penicillin and the other antibiotics.

The study of the mechanisms of immunity to disease generated a wealth of information about vaccines and vaccination procedures, but also provided the foundations of modern immunology in its broadest aspects. By the beginning of this century it was known that non-pathogenic substances could act as *antigens* (substances which elicit immune responses), and the discovery by Landsteiner in 1900, of natural antibodies to blood group antigens, led to major advances in our understanding of basic immunological functioning (and to several Nobel Prizes): Porter and Edelman's chemical analysis of antibody structure; Burnet's theory of the selection of antibody-producing cells from a vast, randomly developed repertoire; Medawar's demonstration of the ability of the immune system to learn what is 'self' and 'non- self'; Dausset and Benaceraff's discovery of the genetic basis for 'self/non-self' discrimination; Jerne's theory of the self-regulating nature of immune responses, and Tonegawa's demonstration of the gene rearrangements necessary for antibody synthesis.

The consequences too, of abnormal immunological activity became the focus of attention: in 1902 Portier and Richet, on board Prince Albert of Monaco's yacht (with what appears to have been a menagerie), demonstrated in various animals the harmful effects (hypersensitivity) resulting from immune responses to otherwise harmless substances. It thus became possible to explain human conditions such as hay fever, and later, pathologies caused by immune reactivity directed at self – the autoimmune diseases.

Progress required, and generated, new technology: application of molecular biological techniques to the isolation and synthesis of antigenic molecules for vaccines; the electrical separation of proteins and DNA; Snell's development of inbred animal lines; Yalow's radio-isotopic assay for high sensitivity detection of antigens and antibodies; the production *in vitro* of monoclonal antibodies (antibodies of absolute purity and desired specificity) by Kohler and Milstein, and the suppression of immunity in recipients of grafted tissues.

MODERN IMMUNOLOGY

Clearly the scope of modern immunology goes far beyond an understanding of the defences against infection and the development of vaccines. Numerous illnesses associated with immune dysfunction are now recognised: *hypersensitivity* reactions including hay fever, eczema and asthma; *autoimmune diseases* such as diabetes mellitus and multiple sclerosis; genetic and acquired *immune deficiencies*. It has even been suggested that the process of ageing is a function of inappropriate immune activity. The success of tissue grafting is as much a problem of minimising rejection whilst maintaining some immunity to infection, as it is a problem for the transplant surgeon, and understanding the genetics of tissue transplantation has given rise to a new sub-discipline – *immunogenetics*.

On the positive side, cancer diagnosis and therapy are benefiting from the use of antibody probes for tumour imaging and from manipulations of anti-cancer immune responses with naturally occurring chemicals produced by immune cells. Maternal-fetal interactions often involve the potential risk of immunological rejection of the genetically non-identical fetus, and immune therapies have been introduced in an attempt to reduce the risk of spontaneous abortion in women with unexplained pregnancy failures. Prophylactic immunisation has also eliminated the problem of Rhesus incompatibility – the birth of jaundiced babies whose red cells have been damaged by maternal immune responses.

The use of highly specific antibodies in diagnosis, therapy and research has enormous potential for the targeting of small molecules in, for instance, medicine, agriculture and forensic science. Forensic science, and even anthropology, have benefitted too, from our understanding of the distribution, amongst individuals and in populations, of the white blood cell markers associated with immune reactivity. Molecular biology has provided the opportunity to study in detail the genetic basis of immune reponsiveness and to create transgenic animal models of human diseases. The techniques available to the immunologist today also allow the detection of individual molecules and the isolation of single cells from populations of several hundred thousand.

The principal aim of this chapter is to acquaint the reader with our current understanding of the immune system in its normal and abnormal functioning, and to illustrate the extent to which immunology has become at once the tool of so many sciences and a discipline in its own right. Recent developments in (especially applied) immunology and the technologies which have made these developments possible, are described in order to explain the present status of immunology, its limitations and its possible future directions.

THE IMMUNE SYSTEM

Innate and Acquired Immunity

It is widely, although certainly not unanimously (see page 112), held that the immune system evolved by virtue of the protection it affords against infection and parasitism. In any case, such protection is of paramount importance and justifies the emphasis placed upon it. In this regard, it should be noted that probably all multicellular organisms exhibit *innate defences* which act to reduce the likelihood of infection. These include anatomical (physical) barriers as well as physiological barriers: the skin is an excellent example of both, since intact skin resists penetration and its low pH (its acidity) is inhibitory to most bacterial growth. Other innate defences include cellular 'eating' (phagocytosis) of micro-organisms and the production of toxic substances such as bactericidins. The existence of innate immunity is independent of antigen, but is non-specific and non-adaptive, that is, shows no evidence of discrimination between pathogens, nor is the response enhanced upon subsequent exposures. For these reasons, emphasis is generally placed upon the adaptive or *acquired immune responses* which are described below, and which constitute the core of this essay.

Acquired immunity, which develops in response to antigen, has two important features: antigen *specificity,* in that responses are directed with great precision to particular molecules or parts thereof, and *memory,* in that second and subsequent exposures to the same molecule elicit a more rapid and more efficient response (the secondary response). The specificity of response is critical, for instance, if we are to avoid making immune responses against 'self' molecules which resemble those of pathogens (see page 117). Specificity is not absolute, however, and the phenomenon whereby a similar molecule is recognised by an antibody directed against another, known as cross-reactivity, can work in our favor (it was the basis for Jenner's cowpox/smallpox vaccination) or against us (damage to heart valves in rheumatic fever is caused by antibodies originally directed at streptococci). The memory of past exposures enables us to carry for life, immunity to diseases such as tuberculosis, after a single childhood immunisation.

Cells of the Immune System

The intricate mechanisms of the acquired immune response involve cells and tissues absent in less complex organisms, but utilise elements of innate immunity as well.

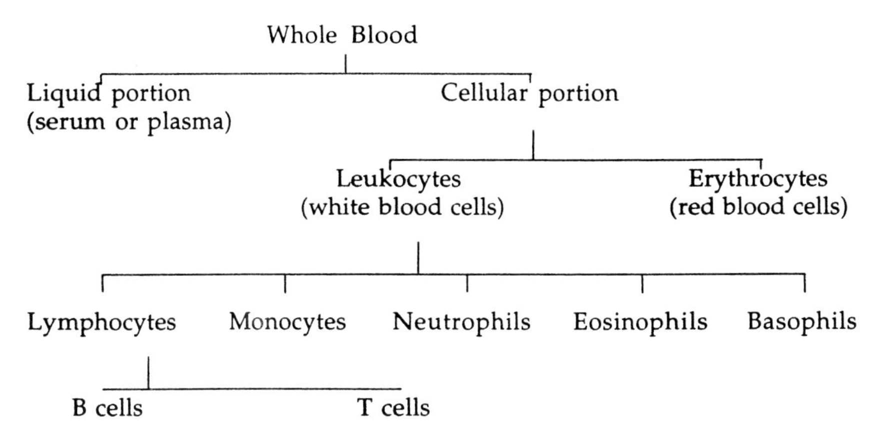

Figure 1. Components of blood

Indeed, the cornerstone of both systems is the phagocyte, some types of which are able to trigger a range of specific immune cells called lymphocytes. Precursor and active phagocytes are found in blood, as monocytes and neutrophils (polymorphonuclear cells or polymorphs) respectively. Also present in blood are two major types of lymphocyte: *B lymphocytes* which have matured in the *B*one marrow where they arose, and *T lymphocytes* which, having left the marrow, have undergone maturation in the *T*hymus (an organ in the upper chest). Collectively, the circulating phagocytes and lymphocytes comprise the blood leukocytes (white blood cells, see Fig. 1), and are responsible for most immunological activity. Two other leukocyte types – eosinophils and basophils – and the non-circulating mast cells, not strictly members of the specific, acquired response, are also important, and will be discussed later.

Tissues of the Immune System

Like all leukocytes, the principal immune cells (lymphocytes) are produced from precursors ('stem cells') in bone marrow and released into the circulation. T cells then undergo maturation in the thymus before returning to the circulation. B and T cells then move between a number of *secondary* organs and tissues (marrow and thymus are the *primary* immune tissues) illustrated in Figure 2. The spleen is a major site of antibody synthesis (although much of its function is associated with red cell turnover). Large aggregates of lymphoid (here used to mean 'immune') cells occur in sites such as adenoids, tonsils, Peyer's patches (on the gut) and the appendix. More diffuse collections occur in many tissues such as the lungs and the wall of the bladder. Most notable, since their swelling in association with on-going infection is frequently a source of some discomfort, are the lymph nodes (sometimes incorrectly called lymph 'glands'). Lymph nodes occur as cell-dense 'filters' along a network of lymphatic vessels which parallel the arterial-venous circulation and collect blood plasma which has leaked into tissues, returning it eventually to the venous system.

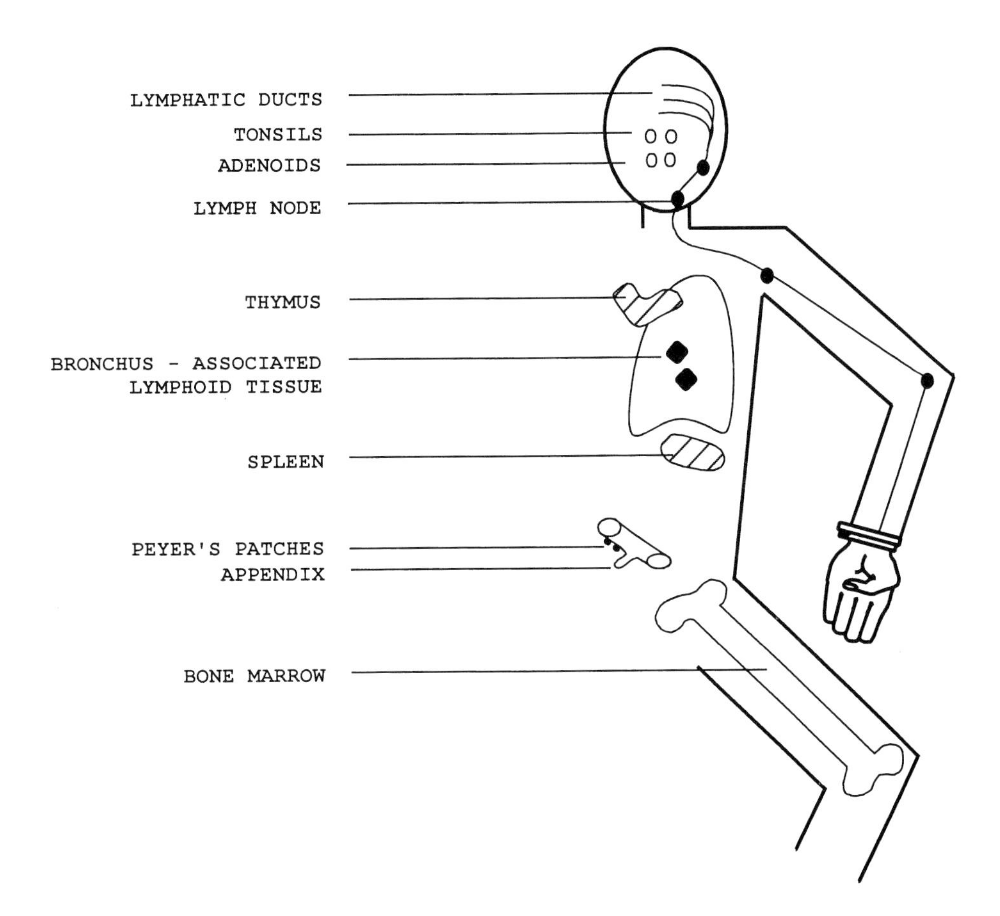

Figure 2. Organs and tissues of the immune system.

[The lymphatics, without the benefit of the heart to pump the lymph (tissue fluid), rely on neighboring muscle contraction and a set of valves to move it. Hence, long periods of inactivity, such as sitting in an aeroplane seat, often result in congestion of the lower limbs and feet. On a more clinical level, the lymphatics, or rather their failure, are perhaps best known through the condition of elephantiasis – the gross swelling of the lower half of the body, caused by parasites blocking lymphatic ducts.]

The movement of immune cells around the body (lymphocyte traffic or migration) is not random. Resting lymphocytes (not yet activated) located in various sites, tend to return to those sites if isolated and re-introduced to the circulation, a phenomenon known as *homing*. Lymphocytes isolated from the gut wall, for instance, tend to re-locate there. Cell traffic is reduced in the presence of antigen, and in lymph nodes this probably maximises lymphocyte exposure and activation. Activated cells may temporarily lose their ability to migrate, increasing their contact with antigen, whilst on the other hand memory cells (see page 106) display enhanced migration.

The immune system should be seen then, as a diverse set of organs and tissues, linked by vast numbers of cells travelling through blood and lymph. The movement of those cells will depend upon their maturity or their state of activation; some will continue to recirculate, while others will take up residence in particular tissues. Ultimately their fate will depend on their role in antigen-specific immune responses.

THE IMMUNE RESPONSE

The historical division of immunity into humoral or cellular mechanisms, which are no longer seen as mutually exclusive, still forms a useful basis for discussion of immune processes. The distinction between the two lies in the nature of the final effector: circulating products in the former, but direct cell-to-cell contact in the latter.

Humoral Immunity

Humoral immunity is antibody-mediated, antibodies being the protein products of plasma cells (the final stage of B cell maturation, occurring in secondary lymphoid tissues). Each plasma cell synthesises a series of antibody molecules (classes or isotypes) which together are called immunoglobulins (Ig's). The first, identifiable by its structure, is called IgM, followed in turn by similar, but discernibly different, IgG, IgA and IgE. [Another, IgD, is poorly understood and will not be discussed here.]

What is remarkable, and made possible only by a series of manipulations involving the genetic material which encodes these molecules, is that all of the Ig's produced by a single plasma cell carry the same recognition sites for antigen. Each antibody molecule has two identical antigen binding sites (see Fig. 3) and these appear first on the IgM and later, for instance, on the IgG, of the plasma cell. Thus, a cell producing antibody to measles virus, will make IgM anti-measles and then IgG anti-measles in what is known as the IgM-IgG switch. In previously exposed individuals, this switch occurs more rapidly (the secondary response) as a result of immunological memory. Subsequent changes to produce IgA and IgE also retain the target specificity.

The functions of antibodies depend upon their antigen binding sites and on properties of the non-binding region. The interaction of antibody and antigen is steric (like a lock and key) and goodness of fit varies (explaining, for instance, cross-reactivity). Binding results in aggregation or agglutination of particulate matter, including antigen-bearing cells, and consequently the formation of insoluble complexes. These immune complexes are deposited in tissues and/or are phagocytosed. Binding can also interfere with function, as in the case of antibodies to cilia or other propulsive devices on micro-organisms. Another consequence of antigen binding is the steric alteration to the non-binding region of the antibody, which, in some Ig isotypes, triggers activation ('fixation') of the *complement* system. Complement is a collection of almost thirty serum and membrane proteins which form a regulated enzymatic cascade resulting, eventually, in the formation of an enzyme capable of producing cell membrane damage and target cell death (lysis). The principal functions of antibodies, then, are antigen binding and cell lysis.

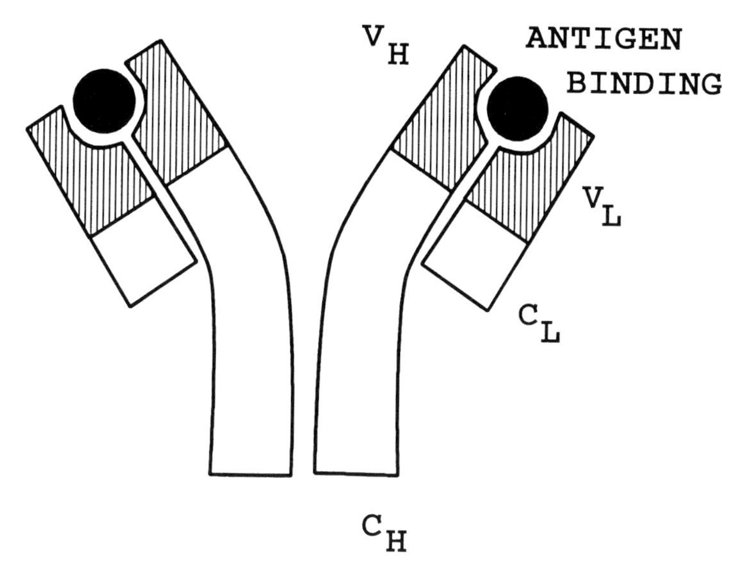

Figure 3. Immunoglobulin (antibody) molecule. Antibodies comprise 2 identical heavy (H) peptide chains and 2 identical light (L) chains, arranged symmetrically as in this highly simplified figure. Each chain consists of a constant (C) region of conserved sequence, and a variable (V) region subject to considerable sequence variation. The variable regions of the light and heavy chains together form the 2 antigen binding sites. The nature of the C_H region determines the Ig isotype (for example, IgM is identified by the presence of so-called μ chain in the C_H region) . The C_H region is changed in the IgM-IgG switch.

Cell-mediated Immunity

Cellular immunity involves the non-specific phagocyte-mediated lysis of pathogens (the killing of bacteria, for example, by neutrophils which ingest and then poison their 'victims' with toxic free radicals) and the lysis of large parasites by eosinophils. It also involves the activities of sub-populations of antigen-specific T lymphocytes, including killer (cytotoxic) T lymphocytes (CTL's) and helper (T_h) T cells. Lysis by CTL's utilises chemicals released upon contact with, for instance, virus-infected host cells, whilst T_h cells secrete different chemicals (cytokines) capable of inducing other lymphocyte and phagocyte responses. Another cell type has emerged in the last twenty years, a part of the innate defence, which has some characteristics of lymphocytes, but also has properties in common with phagocytes. These natural killer (NK) cells are not well understood, but their role in cellular immunity involves both target cell lysis (of cancer cells and virus-infected cells) and the secretion of cytokines.

Finally, mention must be made of another cellular lytic mechanism antibody-dependent cell-mediated cytotoxicity (ADCC). In this process, target cells to which a specific antibody has bound (antibodies not necessarily capable of fixing complement and hence of causing lysis themselves) are killed by any of a number of killer (K) cell types. The antibody here acts only to hold the K cell (usually a phagocyte or sometimes an NK cell) close to the target cell: K cells possess surface receptors for the non-binding region of antibodies. The cytotoxic K cells involved are non-specific, but the specificity of the antibody directs them to specific targets. The *in vivo* role of ADCC is unclear, but it is of interest for several reasons, not least being that it is one of the many examples of a link between humoral and cellular immunity, and between the phylogenetically ancient innate defences and the more 'modern' acquired responses.

Initiation of Acquired Immunity

Lysis of target cells by non-specific effector cells, but involving specific antibody binding (ADCC) is one example of the grey area between humoral and cellular immunity. The initiation of antigen-specific humoral and cellular immune responsiveness provides another: both processes begin with the antigen presenting cell (APC), a specialised phagocyte. All phagocytes ingest material, including whole cells such as bacteria and no longer functional tissue cells, and subject them to chemical digestion. The APC takes this process further by binding small fragments of digested polypeptide (small strings of amino acids) to special molecules and displaying these on the cell surface as antigens (see Fig. 4). Thus, antigen-specific CTL's and T_h cells are able to 'see' the antigens and set in train the appropriate cellular and humoral responses.

Antigen-specific cellular immunity resides with the CTL and the T_h cell. The lytic properties of CTL's allow, for example, virus-infected cells to be destroyed, depriving the virus of host cellular machinery necessary for its replication, and exposing surviving virus particles to phagocytes and antibodies. The role of the T_h cell in this process is to encourage the proliferation of other effector cells and to 'activate' them (enhance their lytic or phagocytic potential) .

A further link between the cellular and humoral arms of the acquired response is also provided by the T_h cell. Whilst some antigens are capable of direct stimulation of B cells (so-called T-independent antigens), most (T-dependent antigens) require the T_h cell as an intermediary between the APC and the B lymphocyte. In this latter case, illustrated in Figure 4, the T_h cell is made aware of the presence of antigen on an APC and in turn alerts an appropriate B cell (the origins of 'appropriate' T and B lymphocytes are explained in the following section). The response of the B cell is to undergo the final stages of maturation, resulting in a plasma cell which secretes large amounts of IgM antibody. The T_h cell has one further function: the switch, from the secretion of IgM to IgG (and later to IgA or IgE) by the plasma cell, is triggered by chemical signals from the T_h cell.

The activation of lymphocytes by APC's both drives the specific immune response (the nature of specificity is described in the following section) and provides immunological memory – the two key features of acquired immunity. When T or B cells are activated, it seems that two cell types result: an effector cell and a memory cell, the former being relatively short-lived, the latter a long-lived 'reserve'. The precise method of memory cell generation is not understood, but such cells retain the capacity to respond in the same specific way as their effector twin, but await

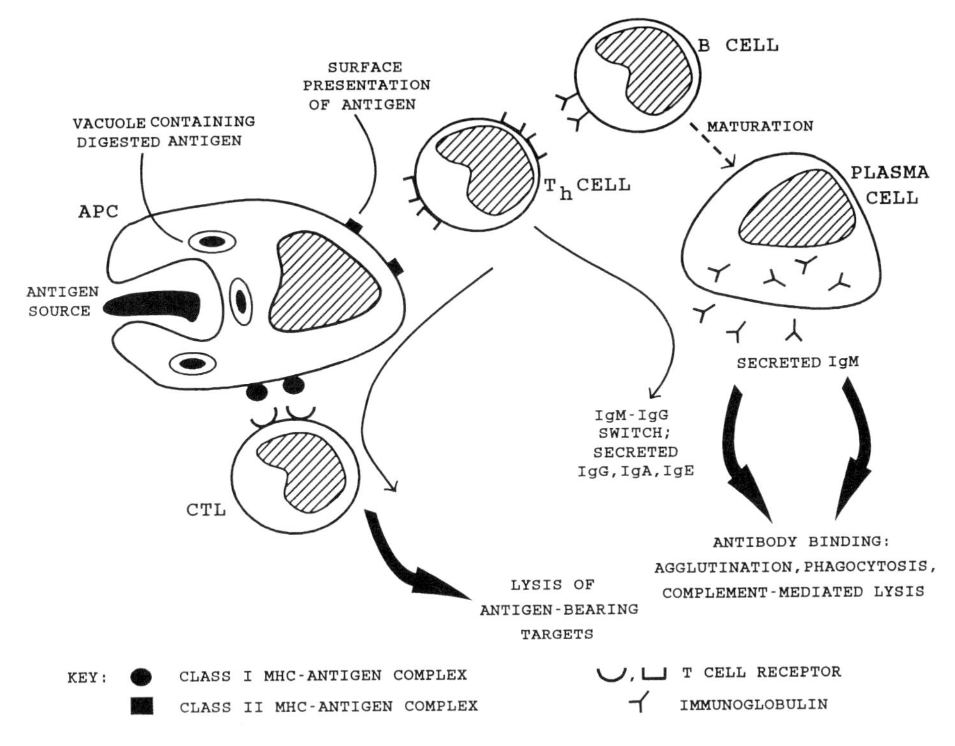

Figure 4. Cellular interactions and the immune response. Foreign material is phagocytosed, digested and small protein fragments presented on the APC surface as antigens. T lymphocytes respond to antigen, effecting cellular cytotoxicity (a function of CTL's) or acting as helper (T_h) cells. The helper cells in turn promote CTL or B cell activity. B cells respond by maturation to plasma cells secreting Ig. The T_h cell also participates in Ig isotype switching. APC antigen presentation occurs in complexes with MHC receptor molecules, class I-antigen signalling CTL and class II-antigen signalling T_h. T cells, in turn, recognise the MHC- antigen through T cell receptors (TCR).

activation by subsequent antigen exposure. They are more sensitive to antigen triggering than the cells from which they arose, and form the basis of the secondary immune response. The reasons why the immunological memory for some antigens is greater than for others, and hence, less frequent immunisation is necessary, are unknown. [Immunity to tetanus, for instance, requires 'boosting' every few years, but, as noted earlier, immunisation against tuberculosis is for life.]

In summary then, acquired immune responses are dependent upon a 'specialised' component of the innate response – the phagocyte-come-APC – and both humoral and cellular immunity are reliant upon the T_h cell. At the effector level, only cell-bound antigen is recognised by CTL's (see below), while both cell-bound and free antigen are recognised and bound by antibodies. Hence, virus-infected cells expressing viral proteins on their surface will be targets for CTL's and antibody, but free virus, in transit to its next host cell, is vulnerable only to antibody.

How is the CTL able to distinguish free virus, or any other cell-free antigen? For that matter, how is it that CTL's and antibodies are directed only at 'non-self'?

The answers to these questions are critical to our immune repertoire, and the nature of injurious anti-self reactivity – autoimmune disease. The answers, perhaps not surprisingly, lie in the genes which encode our immune recognition structures.

GENETICS OF IMMUNE RESPONSES

The Major Histocompatibility Complex

Many immune deficiency disorders have a genetic or heritable basis, attesting to the importance of a large number of genes participating in immune processes. [Acquired immune deficiencies which result from infection, such as the Acquired Immune Deficiency Syndrome (AIDS), may be passed from mothers to offspring, but only by passage of the virus itself.] The influence of genetic background (genotype) on immune responsiveness is far more subtle, however, than can be guessed at by examining genetic immune deficiencies.

In all vertebrates, genetically determined differences between individuals can be demonstrated by the rejection of tissue grafts. Many of these differences are slight, and the rejection responses weak (as in the rejection of skin from inbred male mice by their otherwise genetically identical female littermates). In all species, one set of histo (tissue)-compatibility structures elicits the most powerful rejection responses and the group of genes encoding these structures has been designated the *major histocompatibility complex* (MHC). In humans, because the gene products were first observed on blood leukocytes, the MHC is called the human leukocyte antigen (HLA) system. The importance of the MHC extends far beyond the initiation of graft rejection responses, however, as might be anticipated (since tissue grafting does not occur in nature).

Allogeneic (between members of the same species) and xenogeneic (between members of different species) grafts led to the discovery of the MHC. Independent research into the genetic control of antibody responsiveness in animals also pointed to a set of genes in the same region. The latter, so-called immune response genes, determined the ability, or otherwise, to mount antibody responses to specific antigens. Thus, individuals or groups of individuals with the same genotype, all responded to a particular set of antigens and failed to respond to another set of antigens. Since the graft rejection genes and the immune response genes mapped to the same region – the MHC – they came to be called the MHC class I and class II genes respectively. What proteins do these genes encode? And how do they fit into the processes described above?

The MHC class I genes are expressed in all nucleated cells (all cells except mature erythrocytes in humans) and their products are found on the outer surface of the cell membrane. The human class I genes include three loci, A, B and C which play a critical role in tissue transplantation. Since we receive a set of these genes from each of our parents, and the genes are co-dominantly expressed (there is no masking of one 'recessive' gene variant by another 'dominant' one), we each express two variants (alleles) of the A gene, two of the B and two of the C. In the population at large, many alleles exist for each of these genes (approximately 50 B alleles are known, for example) and such a system is said to be polymorphic. The likelihood of locating a tissue donor with a matched set of such alleles (a set of alleles on one chromosome is called a haplotype), or more importantly, two such matched sets, is of course, very low: over one million randomly chosen donors would be necessary to include a 'perfect match'. How fortunate then, that most of us do not require organ transplants! The haplotypes we carry are, however, important on a day-to-day basis.

The MHC class I molecules act as receptors that bind small proteins inside the cell and display them on the cell surface. Any protein small enough (less than ten amino acids) and able to fit the molecular shape of the particular class I receptor, can be bound and displayed. Most such proteins will be 'self' and be representative of the cell's protein synthesis, but in, for instance, virus-infected cells, the major proteins available will be viral in origin. It is therefore possible for a cell to signal its infected state to virus-specific CTL's and to antibodies. The virus-specific CTL is first alerted to the virus when an APC, having 'processed' the virus, presents the degraded antigen, complexed to MHC class I, on its surface. The CTL lyses antigen-positive cells, but is only able to recognise the antigen in its MHC-complexed form, hence it is not 'distracted' by free antigen. [antibodies are able to recognise and bind free antigen as well as the exposed portion of antigen in the MHC-antigen complex].

The MHC class II molecule is also a receptor binding protein, but the genes and their products are expressed only in cells with an APC function. As with the class I genes, the class II region contains several loci, in humans called DP, DQ and DR, and each is polymorphic and co-dominantly expressed. Their limited tissue distribution, however, renders them much less important in the context of transplantation. Instead, their ability to bind antigen, and their location on the surface membrane of APC's, enables them to signal T_h cells, in turn driving both cellular and humoral immunity. Again, the binding of antigen to MHC receptor is steric, and only antigens of appropriate size (probably similar to those bound to class I) and shape will 'fit' and be displayed by any particular receptor.

The particular class I and class II MHC haplotypes then, determine the repertoire of antigenic peptides which can be bound and to which CTL's and T_h cells respectively can respond. Hence, indirectly, they determine the repertoire of pathogens (micro-organisms, parasites) to which immunity can be attained. As a result of natural selection, populations probably exhibit an array of MHC molecules which reflect their infectious history, although no individual is likely to carry all of these. A major selective force for the maintenance of what is the most polymorphic gene system yet described, is thought by many to be the need, at the population level, to ensure some survivors of any potential epidemic. It is argued that the vast array of MHC alleles guarantees that no infectious agent can successfully avoid stimulating immune responses in all members of a host population [another explanation for the polymorphism is discussed later – see page 120].

Whatever the limitations placed on the immune repertoire by the finite number of MHC alleles carried by an individual, because of the enormous number of peptides which 'fit' each one, the immunological capability is astonishingly high. How, then, is an individual able to generate a specific response to each of the millions (perhaps billions) of antigens which the MHC molecules can bind? This too is a genetic problem, but its solution is very different.

Lymphocyte Receptors

Lymphocytes (CTL's, T_h and B cells) all carry on their surfaces, receptor molecules which specifically bind an MHC-antigen complex (and, in the case of B cells, antigen alone). The total genome is not of sufficient size that a gene for each specificity could be present. Instead, the genes for the *T cell receptor* (*TCR*) and immunoglobulin (which, in the form of early IgM, is membrane bound and functions as the B cell receptor) are limited in number, but undergo remarkable rearrangements. These rearrangements, and the extraordinary number of new variants created by mutational 'hot spots', result in the generation of many millions of different genes and in turn, different molecules.

Briefly, the genes coding for TCR (there are four sub-units, a, b, c and d) and Ig molecules each comprise one or more constant regions which characterise the molecular type (TCR alpha, for instance, or IgM) and a large number (probably several hundred) of variable regions (see Fig. 5). The variable region repertoire for Ig's is further expanded by an unusually high number of mutations (alterations to the original genetic message) occurring throughout the life of the individual. As a result, no two lymphocytes will necessarily carry the germ-line (inherited) set, nor will they necessarily carry the same sets as each other. In order to produce a particular receptor type, bearing a particular antigen specificity, the appropriate constant region (eg TCR) is 'selected' and joined to the requisite variable region (eg recognising measles virus), creating a new gene (Fig. 5). Variable region 'choice' or 'selection' is probably random (as is its generation by mutation in Ig's), but occurs independently in each T or B cell.

Since billions of these cells are produced in bone marrow every day, similar numbers of antigen-specific receptors must also be generated. Most of these cells will carry specificities which are not needed, in the absence of the particular antigen, and the cells will undergo a process of 'cell suicide' called apoptosis.

This may seem wasteful, and is indeed a great drain on energy and resources, but it allows that a useful specificity will appear and be available when needed: the same specificity will occur (by chance), again and again over time. Thus we have the potential to respond to antigens which we have never seen and which might not even exist at present, and all with only a modest investment in genetic material.

The basic picture is now complete. T lymphocytes (CTL's and T_h cells) are constantly generated and carry novel surface receptors. In the absence of antigens, or more correctly, MHC-antigen complexes, which match the receptors, the cells apoptose. If, on the other hand, they encounter an APC on which the appropriate MHC-antigen is presented, they are activated and an immune response proceeds. B cells, using their novel Ig as a surface receptor, similarly 'suicide' if no antigen match is found, or else complete their maturation and secrete the Ig in large quantities [antibody is clearly 'selected' by antigen, an idea proposed by Ehrlich in 1894 and revived in the 1950's by Jerne and more elegantly, by Burnet, who believed

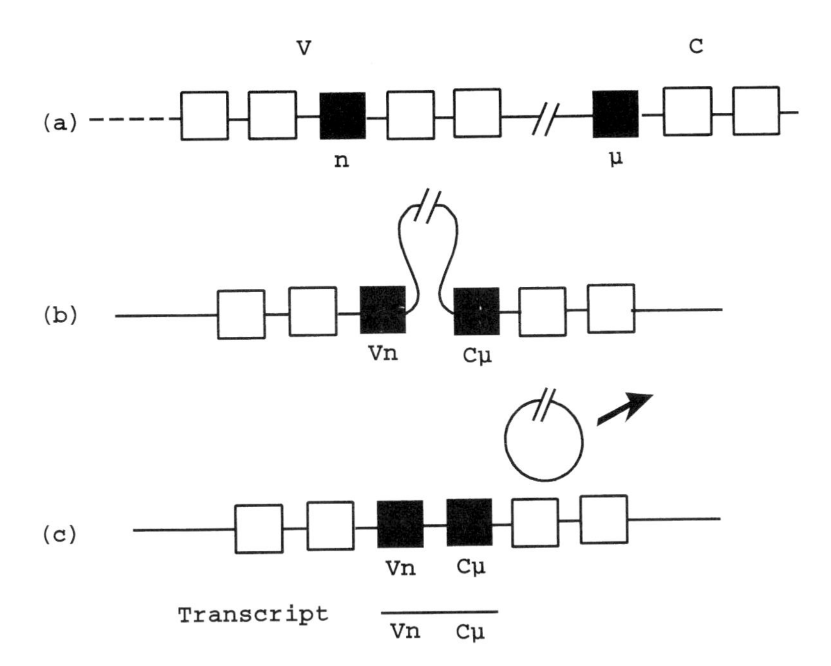

Figure 5. Immunoglogulin and T cell receptor genes. In germ-line configuration (a) many hundreds of variable (V) region genes are separated by some distance (and some additional regions not discussed here) from a few constant (C) region genes. In the example above, the creation of an IgM heavy chain molecule is shown, using the μ constant region gene, and, bearing the particular antigen specificity determined by gene n. The intevening DNA forms a loop (b) bringing the requisite genes together and eventually (c) the loop is excised and the DNA re-joined to form a new gene from which a complementary message (transcript) is formed. Similar rearrangements occur in TCR genes.

his clonal *selection theory* to be more deserving of a Nobel Prize than the work on tolerance (below), for which he actually received it].

Recognition of 'Self' and 'Non-Self'

It is inevitable that, in the random generation of antigen-specific TCR and Ig molecules, many will appear which will recognise and bind 'self' peptides. All uninfected host cells too, will bear self peptides in their MHC class I molecules, and APC which have scavenged dead or dying host cells, will display processed self peptides in both class I and class II molecules. What then, prevents immune reactivity against self antigens? [This is not a trivial question, since one obvious consequence of self-reactivity is auto-immune disease, of which various forms of arthritis and diabetes are common examples (see page 112).]

Non-identical twin calves frequently share a placental circulation and consequently share blood cell precursors during fetal life. This observation, made in 1945 by Owen, led Burnet to correctly postulate that exposure to self peptides during fetal and early neonatal life allowed the developing immune system to recognise self. In

1953, Medawar verified this idea by showing that skin grafts could be successfully exchanged between such twin cattle. It would appear that the developing immune system 'learns' what is self, and responses can only be elicited to antigens which are encountered (for the first time) after this period. These experiments resulted in a Nobel Prize for Burnet and Medawar and formed the basis for our present understanding of the phenomenon of *tolerance*.

The normal state of unresponsiveness to self is called self-tolerance. The means by which this is achieved remain unclear, but two mechanisms seem likely. Within the thymus; potentially autoreactive T cells and their progeny (clones) are probably eliminated – a process called clonal deletion – when the cells are immature. This involves the recognition of self via the MHC-self peptide complexes displayed by the non-lymphoid cells of the thymus (and the induced apoptosis of T cells directed at them). In the periphery; in secondary lymphoid tissues, for instance, mature autoreactive T cells undergo functional inactivation, by an as yet unknown mechanism which prevents them from responding to normal signals. Autoreactive B cells too can be eliminated or inactivated, but again the precise mechanisms are not understood.

IMMUNE DYSFUNCTION

Autoimmunity

Under normal circumstances then, the maintenance of self-tolerance prevents the activation or functioning of self-reactive lymphocytes. As we grow older, however, self-tolerance breaks down, in ways yet to be understood, and reactivity to self molecules becomes apparent. This autoimmune reactivity is not necessarily pathogenic and many of us probably produce, for instance, antibodies which recognise intracellular proteins without ill effect (antibodies do not cross the membranes of cells). The breaking of tolerance to cell surface molecules or extracellular protein presents a different picture, and usually results in autoimmune disease. One form of diabetes, for example, results from auto-reactive T cells directed at the pancreatic cells producing insulin.

Autoimmunity results from the breakdown of tolerance, and many common clinical conditions exhibit, to a greater or lesser extent, autoreactivity, including rheumatoid arthritis, multiple sclerosis and even some forms of hepatitis. The basis for loss of tolerance is unclear, but it was observed earlier that 'self' is learnt during development of the immune system and so tolerance will not exist for molecules unavailable or not seen during this period. This has at least one advantage, however, as cancer cells often express proteins otherwise restricted to very early development (too early to be 'screened' during the immune system's early learning programme). Consequently, the lack of tolerance to cells expressing these proteins results in a degree of anti-cancer immunity and is the basis for new strategies aimed at immunotherapy. [Burnet and others have even argued that surveillance against abnormal (cancerous) cells has been the real driving force behind the evolution of acquired immunity.]

Hypersensitivity

One consequence of inappropriate (in this case, anti-self) reactivity is autoimmune disease. Another result of inappropriate or exaggerated reactivity with potentially harmful, sometimes even fatal, effects is hypersensitivity, of which allergy is the best known example. Hypersensitivity involves no unique immunological features (that is, none that has not been shown to occur in the eradication of infectious or parasitic organisms) but is associated with tissue damage. In allergic reactions, individuals with a genetic predisposition, produce very high levels of allergen (allergic antigen)-specific IgE which causes the release of various chemicals such as histamine from basophils and mast cells. These chemicals are of considerable value in the response to parasite infestation, but in allergic individuals result in asthma, hay fever and eczema. They act to increase fluid 'leakage' from blood into tissues and cause swelling and itching, typically seen in positive skin test responses to allergens such as pollens, cat dander and house dust mites. In extreme cases, the chemical response to inhaled allergens or, for instance, to bee stings, is of such magnitude that breathing is fatally impaired.

Other types of hypersensitivity involve exaggerated IgM and IgG responses which trigger localised or systemic (general) tissue damage following the deposition of immune (antigen-antibody) complexes and the activation of complement. In addition to the lysis of bystander (host) cells, complement components are able also to stimulate inappropriate phagocytosis by neutrophils in organs such as the kidney. Autoimmune antibodies are also thought to be able to trigger tissue damage by the binding of K cells, resulting in lysis by ADCC.

A familiar example of yet another type of hypersensitivity is contact dermatitis, which is delayed (maximal response takes about 48 hours) and mediated by T lymphocytes and phagocytes (especially monocytes). A clinically useful form of this response is the basis of the Mantoux test for immunity (successful vaccination) against tuberculosis. On the other hand, chronic tuberculosis and a number of other mycobacterial, protozoan and fungal infections are often accompanied by both protective immune reactivity and pathogenic delayed hypersensitivity.

Immune Deficiency

Failures of the immune system to respond adequately, result in frequent, prolonged and severe infections, often caused by organisms of otherwise low pathogenicity. Such failure may result from a primary deficiency – hereditary or developmental – and can affect the bone marrow stem cells, T or B lymphocytes, phagocytic cells or the complement system. Some are highly specific and may even be of little consequence, a common example being the absence of the IgA isotype (a condition which has a frequency of 1 in 600). Others, fortunately relatively rare, affect major components of immunity with disastrous consequences. The famous "boy in the bubble", whose severe combined immune deficiency (SCID) syndrome deprived him of cellular and humoral immunity, survived life-threatening infection only by complete isolation [he died, aged 12, from complications following a bone marrow transplant aimed at providing him with a new and functional immune system].

Many immune deficiencies are secondary to other conditions such as infection, malnutrition, burns or drugs. The best known of these, the Acquired Immune Deficiency Syndrome (AIDS), has taken on great significance since its emergence in 1981. Caused by a retrovirus now called the Human Immune Deficiency Virus (HIV), recent evidence suggests its presence in Africa since the 1950s. Ironically, a major target for HIV infection is the T_h cell in which the viral genetic material becomes integrated with the host's before replicating and lysing the lymphocyte. The virus is able also to reside safely within phagocytic cells and to bring about the deaths of other, uninfected, immune cells, resulting in eventually fatal infection due to the absence of adequate cellular or humoral immunity.

Finally, brief consideration must be given to the perturbations in immune status, both stimulatory and inhibitory, associated with a variety of environmental and lifestyle factors. Minor diurnal variations in, for instance, circulating leukocyte numbers, and the disastrous immune suppression induced by chronic malnutrition, have long been recognised. Only recently, however, have the effects of moderate and severe exercise, smoking, alcohol and other drug use, ultraviolet light and altered sleep patterns, emerged. Some of these effects are probably the result of stress-related hormone activity, and a new science – neuroimmunology – has been spawned to investigate the link between the nervous, endocrine and immune systems. The immunological impact of physical and emotional stress can be seen in examples as disparate as the effects of reduced gravitational forces experienced by astronauts and the bereavement associated with the death of a close friend.

How is it that so much is known of the workings of the immune system? (Perhaps the question should be: "What do we *not* know ?") And how have immunological approaches come to be used in answering a wide range of non-immunological questions? And where is immunology headed? The following sections briefly address these issues.

TECHNOLOGICAL DEVELOPMENTS

The biological sciences have undergone a technical revolution over the past twenty years, and immunology, like many disciplines, has reaped the benefits of major advances in molecular biology (recombinant DNA technology or 'genetic engineering'). Immunology has advanced too, as a result of specific developments, especially in antibody synthesis and detection.

Advances in Molecular Biology

Manipulation of the genetic material, deoxyribose nucleic acid (DNA), became possible in the period 1967–1977 when the enzymes and techniques were developed which allowed DNA to be cut, re-joined, replicated (cloned) and its composition rapidly sequenced. Since that time, further advances have occurred at a stunning rate. Genomic DNA libraries can be created, containing all of the genes of an organism. Gene cloning in bacteria, itself a major innovation, has largely been superseded by the polymerase chain reaction (PCR) which will produce over a million copies from a single fragment in two hours! DNA fragments can be separated by applying an electric field across a porous medium in a process called electrophoresis, and the identity of fragments can be determined using complementary DNA, or ribose nucleic acid (RNA), in what are known as Southern or Northern blotting respectively. The precise order of the nucleic acid bases in a fragment can also be determined rapidly by automated DNA sequencing.

The applications of these technologies extend far beyond the scope of this review, but some examples will demonstrate their power. Probably the best known application of genetic engineering is the transgenic animal, in which all cells carry altered or additional genetic material, as a result of manipulation of the egg or early embryo. The normal role of specific genes or their products can be determined using such animals in which the gene of interest is inactivated or 'knocked out'. IgE knock-out mice have been used, for instance, to assess the role of IgE in inflammatory processes, and in other knock-out mice, the function of accessory proteins in MHC class I receptors has been investigated. Neither of these experiments had been possible prior to the creation of transgenic animals.

As a second example of the application of these techniques, consider the use of DNA sequencing in immunogenetics. The MHC class II molecules of a strain of diabetic mice have been implicated in the aetiology of the disease. Sequencing the genes revealed a receptor type unique to the diabetic strain. Moreover, sequencing of the equivalent genes in human subjects demonstrated the same receptor amongst diabetic individuals. From these data may emerge some understanding of the self protein presented and targeted in this autoimmune disease and perhaps even a diagnostic tool, neither of which was possible using conventional physico-chemical or immunological procedures.

The combination of classical genetics and molecular biology is rapidly expanding our ability to locate (map) genes of interest, including those associated with immunological phenomena. Already possible too, but in its infancy, is the application of molecular techniques to gene therapy (the replacement of 'faulty' genes) and the first disease for which this has been attempted is an immune deficiency similar to that which afflicted the "boy in the bubble".

Technical Advances In Immunology

Undoubtedly the most significant recent advance in immunological technology has been the ability to produce cellular antibody 'factories' by fusing immortalised cancer cells with plasma cells synthesising a desired antibody. Kohler and Milstein shared (with Jerne) the 1984 Nobel Prize for this 'hybridoma' technology, since the production of virtually unlimited amounts of single specificity *monoclonal antibody* has revolutionised immunology and provided research tools for many other disciplines. Conventional (polyclonal) antisera raised in animals are naturally contaminated with antibodies directed at many of the other antigens to which the animal was exposed. Monoclonal antibodies, by contrast, have but one target specificity. [The clinical use of xenogeneic (usually mouse) monoclonal antibodies for imaging or therapy is hindered, unfortunately, by the host immune response to the mouse Ig constant region. DNA technology is overcoming this problem by replacing the non-binding region with a human equivalent. At least until the production of human hybridomas, mouse antigen-specific binding regions can be generated on otherwise human antibodies.]

Techniques for the detection and quantitation of antibodies have also shown remarkable progress. The radio-isotopic and chemical tagging of antibodies allows their visualisation in a range of assay types, of which the enzyme-linked immunosorbent assay (ELISA) is perhaps the best known. Like many systems, the ELISA utilises tagged secondary (marked) antibodies which bind to primary (test) antibodies, which in turn are bound to specific antigen, in a 'sandwich' technique. Similarly, this approach is used in conjunction with electrophoretic protein separation in what is known as Western blotting. Here separated proteins are transferred from their electrophoretic medium (a soft gel) to more durable nylon or acetate and 'probed' with specific antibodies. It is possible to detect nanogram quantities of specific antibody using ELISA and purified antigen, and to demonstrate microgram quantities of antigen in a protein mixture using electrophoresis and Western blotting.

Although this review is not intended to be exhaustive, it would be remiss to ignore the most significant advance in cell identification and isolation – flow cytometry. A combination of highly developed laser optical techniques and tagged monoclonal antibodies directed at cell-specific marker proteins, now allows rare cells (one in several hundred thousand) to be counted, or mixtures of cell types to be analysed. Inclusion of electrical field generators in the system allows those cells (even the rare ones) to be isolated and (with or without further cloning) to be studied. [The ability to maintain cell lines in culture and the elaboration (and now the synthesis) of many of the growth factors necessary for specific cell maturation or proliferation, have also played an essential part in the analysis of cell function.]

What then, are some of the most recent developments in our understanding of immune processes which have resulted from these technological advances?

DEVELOPMENTS IN IMMUNOLOGY

Much of the mystery surrounding antibody function dissipated when Porter and Edelman demonstrated the structure of the molecule (Fig. 3). But how was specificity maintained when a plasma cell underwent the IgM-IgG switch and changed isotypes? And how could an individual produce more antibody types than there were genes in the entire genome? The answers to these questions were provided by Tonegawa's revelation (1987) that a finite set of genes could be shuffled and reassembled to create a vast repertoire of new genes (Fig. 5). It is now possible to study the genes and enzymes *controlling* these Ig gene rearrangements and already at least two have been identified. Defects in these controlling genes result in faulty enzymes and inappropriate Ig gene constructs, and are clinically manifest as, for instance, some forms of SCID. The current emphasis then, is on an understanding of the regulation of gene rearrangement.

The generation of the repertoire of T cell receptors, on the other hand, has only been revealed in the past decade, partly because the TCR is always membrane bound and not secreted. Perhaps not surprisingly, the TCR proteins closely resemble Ig's, and the gene structure and rearrangement are also similar. The potential diversity of TCR's is, however, much greater than even that of the Ig repertoire (approximately 10^{15} or 10^{18} compared with 10^{11}) although the demands on genetic material are no greater, and mutation of the TCR variable regions seems to be minimal. Presently, the precise functions of the different TCR types are under investigation, and much effort is being invested in elucidating the complex accessory molecules now known to be involved in TCR-MHC binding.

The discovery, in the 1970's, that CTL's recognised only MHC-complexed antigens, led to intensive study of the relationship between the MHC receptor and its bound peptide (the optimal peptide size, for instance, is different for class I and class II molecules). Most recently, the 'processing' of antigen – the breakdown of larger peptides and the complexing with MHC – and the externalising of the complexes, have begun to be clarified.

Conventional antigens stimulate a very small proportion of the total T cell population (probably one in ten thousand at best), but a group of substances called superantigens, first described in 1990, have the ability to stimulate up to 20 per cent of T cells in an individual. These include a number of bacterial toxins, and they act by binding together APC class II molecules and T_h cell TCR's without the need for antigen-specific matching (the superantigen binds both receptors *outside* the antigen binding regions). The superantigens have assumed considerable importance through their association with conditions such as food poisoning and toxic shock syndrome (probably a result of abnormal cytokine production). They are also under suspicion for their possible role in autoimmune disease, since self-reactive T cells could be activated by the non-specific stimulation, and superantigen injection induces autoimmunity in some animal models. The search is now on for superantigens or their sources in a number of autoimmune conditions.

Autoimmune diseases affect perhaps 7% of the population and are of considerable clinical importance. Animal models of several human autoimmune diseases are now able to be exploited in the search for the mechanism(s). The results to date implicate a number of different causes. The exposure of the immune system to antigens normally isolated from immune cells (in so-called 'immunologically privileged' sites such as the brain, anterior chamber of the eye and the testis) can result in autoimmune reaction. Sperm antigen release following vasectomy, for instance, can lead to autoantibody production. The similiarity between self antigens and those of infectious organisms (cross-reactivity or molecular mimicry) can also lead to autoimmunity. Many viruses, for example, express antigens similar to those of the insulating protein myelin which surrounds nerve cells, and autoimmune encephalitis can follow viral infection. In other autoimmune conditions, cytokine imbalance, inappropriate MHC class II expression and T_h cell dysfunction are amongst the underlying causes.

Immune responses to self are usually curtailed by the deletion or inactivation of appropriate T and B cells. The precise mechanisms by which this is achieved are uncertain, but the use of transgenic animals is providing a new avenue of research. A 'double transgenic' mouse, for example, has been produced which expresses a chicken protein and large amounts of antibody to the chicken protein. In this system, tolerance is dependent on the *amount* of antigen produced ('seen') during early development. Similar experiments are in progress to examine tolerance to exogenous antigens and the possible breakdown of the process in autoimmunity.

Leukocytes enter the tissues by passing between the cells lining the blood vessels. These cells express cell adhesion molecules (CAM's) and the leukocytes bear complementary receptors which allow them to bind to the vascular tissue. Several families of CAM's are now identified, many cloned and sequenced, and their structures appear to be similar to those of Ig's and TCR's. Some are responsible for lymphocyte homing, being present only on the vasculature of specific tissues (Peyer's patches, for example). More importantly, CAM's are necessary not just in assisting leukocyte movement, but in cell-cell interactions such as T_h-APC and CTL-target. Moreover, abnormal CAM's are associated with increased bacterial infection and are likely to prove of clinical significance.

Immunological memory continues to defy explanation, but recent observations of the distribution of homing receptor CAM's explain the pattern of memory cell circulation. Memory T cells lack homing receptors for secondary lymphoid tissues, but move in tissues such as lung and skin by virtue of other receptors on their surfaces.

As a final example of the 'state of the art', let us consider the cytokines, the chemical mediators of immune cell-to-cell communication. Gene cloning and sequencing have revealed a number of multifunctional proteins, some stimulatory and others inhibitory to immune responses. Abnormal functioning of these molecules or their cellular receptors, is associated with diseases such as bacterial toxic shock and some cancers. A fuller understanding of their activities holds promise for therapeutic immune regulation in, for instance, cancer, allergies and tissue grafting.

APPLIED IMMUNOLOGY

Earlier in this chapter, the scope of modern immunology was outlined. In light of the subsequent discussions, in particular the recent advances in our understanding of immune processes and the technological developments, it seems appropriate now to explain more fully the present applications of immunology, before closing with some comments on its future.

The roots of immunology are to be found in immunisation, just as the evolution of the immune system probably proceeded in parallel with the evolution of infectious organisms. A major application of immunology today still lies in the understanding of host responses to such organisms and in developing vaccines. Present strategies emphasise molecular vaccines – genetically engineered molecules identical to those of the infectious agent and which are likely to trigger successful immunity. There is still a need, however, for much basic research into the nature of host responses and the ways in which (especially parasitic) organisms evade immune defences (one such being the ability to change surface antigens).

Immune dysfunctions are of great clinical significance, varying in their impact from mildly annoying (hayfever, for instance) to severely reducing quality of life (such as crippling arthritis or infertility) through to life-threatening (AIDS or autoimmune liver disease, for example). As is the case for research into the immunology of infection, here too are opportunities for basic and applied efforts, particularly in the design of therapeutic strategies. In this regard, two current approaches to hypersensitivities are 'desensitization' with crude (or genetically cloned and purified) allergen, and the development of drugs which interfere with various steps in the IgE–mast cell/basophil pathway. Future treatment for autoimmune disease (most present treatment is simply a systemic immune suppression, which has obvious contra-indications) is likely to rest with suppression of the specific autoimmune response, using 'designer' peptides which compete at the MHC receptor or TCR level with the autoantigen. Immune deficiencies are already the target for gene therapy, since many are manifest in bone marrow stem cells which can be accessed readily, manipulated and replaced. For other such conditions, a vast armoury of natural and synthetic immune mediators such as cytokines is being developed.

At the forefront of applied immunology since the first successful kidney graft in 1954, (for which Murray shared the 1990 Nobel Prize), has been organ or tissue transplantation. Two immunological problems stand in the way of success: 'matching' the patient and donor at the MHC/HLA gene loci, and suppressing the patient's immune response against the transplant without totally compromising resistance to infection. Molecular biology is improving the speed and accuracy of HLA typing, and immune suppression can be achieved with radiation, drugs or even monoclonal antibodies. More recent experimental approaches have instead, utilised the induction of tolerance to grafted tisue (by introducing antigens into the thymus), or by taking advantage of the immunologically privileged sites (grafting fetal brain tissue into the brains of Parkinson's disease patients). Artificial privileged sites (acrylic vessels) are already reported in experimental systems and high hopes are held for them.

Forensic science and anthropology have also seen the benefits of HLA typing. The complex genetics of HLA inheritance and expression, in combination with developments in molecular biology, have made HLA a prime tool in the identification of individuals. High levels of exclusion and the ability to type from small samples, using PCR, have greatly facilitated tissue typing. The precision and speed of molecular typing has meant that population movements, genetic admixtures, and even the time frame are now amenable to analysis using single HLA alleles and HLA haplotypes (the 'decay' or breakdown of which is slow but calculable).

The application of antibodies, especially monoclonal antibodies, in the identification or localization of specific molecules, has revolutionized immunology and provided a research tool for many other disciplines. For example, in a 'role reversal', antibodies are used sometimes to screen gene banks where the molecular biology has allowed the generation of protein products. In clinical medicine, antibodies to tumour antigens have been used to target anti-tumour agents or to localise secondary tumours which are undetectable by more conventional imaging. In 1990, antibody function was extended once again, this time as a catalyst for chemical reaction, when a million-fold increase in reaction rate was reported for the synthesis of a particular carbon compound using an artificial antibody to bind a transition state or intermediate. These artificial enzymes (or 'abzymes') have the potential to take over the catalytic market.

The semi-allogeneic fetus (the offspring of unrelated parents) half of whose genetic make-up derives from its father, presents a paradox to reproductive immunology. Why do the paternally-derived antigens, especially those of the HLA system, not induce a rejection response from the mother? In fact, it seems that they do, since multiparous women exhibit antibody reactivity to their partners' HLA, but this does not appear to result in rejection. Some people believe, however, that a proportion of unexplained spontaneous abortions are the consequence of immune rejection. What is paradoxical, is the argument that the spontaneous abortions are associated with *similarities*, not disimilarities, between partners (an observation akin to that of genetic inbreeding). Thus, it seems, that immunologists and geneticists would assume opposite requirements for a successful pregnancy. This is one area, along with immune-mediated infertility, where immune therapy and reproductive medicine meet (spontaneous abortion is sometimes treated by immunization). [An alternative to the view that the extreme polymorphism of the MHC is a result of co-evolution with parasites and other infectious organisms, follows from the data on spontaneous abortion. Supporters of the alternative, argue that the MHC arose as a mechanism to maximise outbreeding. The ideas are complex and the data controversial, but the rationale for immunotherapy lies within this framework.]

Finally, in considering the broad applications of immunology, mention should be made of the efforts to understand the relationship between stress and immune reactivity. Lymphocytes have been shown to express receptors for various neurological transmitters and many stressful situations result in immune suppression. The identification of individuals at risk is sometimes difficult, but an example is the possible immunological cause for frequent illness in shift workers, whose day-night cycles are disturbed. Practitioners of 'neuroimmunology' will doubtless continue to discover immunological disturbances in many of us going about our normal lives.

FUTURE DIRECTIONS

Clearly immunology, as a discipline and in its broadest applications, impinges on a large number of basic and medical sciences. But what of its future? What questions remain and which of them will attract the support necessary for a solution? The answers, to a large extent, are evident in the preceding section, but the following represent the images in the author's crystal ball.

Arguably the most important theoretical issues for the immunologist concern the regulation of immune activity, and the basis for memory and for tolerance. Jerne shared the 1984 Nobel Prize for his theoretical explanations of the self-regulating nature of immune responses, but the ideas and for instance, the vaccines based upon them, remain controversial. Understanding memory and tolerance will also be intellectually satisfying as well as clinically applicable in immunisation and tissue grafting respectively.

The 'antibody revolution', begun with monoclonals, will doubtless continue as more and more problems of molecular identification or localisation arise and new generations of antibodies appear. Driven by molecular biology, hybrid (mouse-human) antibodies are already available, human monoclonal antibodies await release, and totally engineered antibodies made by DNA transfer into various micro-organisms have been reported.

Autoimmune disease will surely occupy the ageing human population to an ever increasing extent. The spectrum of autoimmune tissue damage continues to grow, as do the apparent mechanisms responsible. Common debilitating conditions such as rheumatoid arthritis and insulin-dependent diabetes mellitus, as well as, for instance, autoimmune infertility, severely reduce the quality of life. Other autoimmune disorders may be fatal, and of great interest here, are the recent claims that the manifestations of AIDS may be autoimmune in nature.

As the second most common cause of death in western countries, and rapidly increasing in developing nations, cancer has, and will continue to be, a major target of medical research. Immune surveillance against cancer is still a moot point, but immunological contributions to diagnosis and to therapy will undoubtedly continue. Vaccines hold out hope, although the heterogeneity of cancer types remains a barrier to a 'universal' cancer cure, especially in the face of failure to identify unequivocally tumour-specific antigens.

Bringing the discussion full circle, it seems fitting to conclude on the topic of infection and vaccines. More than 650 million people world wide are infected at any one time by protozoan and helminthic parasites, and many millions of them will die as a result. Most of these are in countries where resources are poor and vaccination is the only workable strategy. But despite the advances in treatment and prevention of other human infections and many animal parasites, no successful vaccine exists for any human parasitic disease. In part, this reflects the strategies which parasites have evolved for immunological 'escape', but there are still very large gaps in our understanding of the immunological responses to parasites, and even, it seems, whether in all such cases, the relationship is parasitic or symbiotic.

As for other infectious agents, communicable diseases remain major concerns throughout the world as new infections continue to emerge. AIDS is perhaps the most notorious, but Legionnaire's disease, Lyme disease, the Lantavirus pulmonary syndrome reported in 1993 in the USA, and infections induced by drug-resistant strains of previously controlled pathogens such as *E. coli*, *Mycobacterium tuberculi* and other bacteria threaten to become epidemic (some 20,000 cases of drug resistant tuberculosis have been reported in New York, for instance). The task is daunting, but it seems particularly apt to return here to the case of smallpox, described at the beginning of the chapter, and for which the first recorded immunisations were attempted. As a result of probably the most successful public health exercise ever attempted on this scale, the last naturally occurring case in the world of the great scourge smallpox was reported in Somalia, on October 26, 1977, and today the disease has ceased to exist on the planet. Surely this is cause for cautious optimism.

Acknowledgements

The author is indebted to Helen Skene for her deciphering skills, patience and effort in the typing of this manuscript. Thanks also to Larisa Chisholm for helpful discussion, and Pam Bourton for translating my diagrams.

Further Reading

Ada, G. L. and Nossal G. 1987. The clonal selection theory. *Sci. Am.* **257** (2): 50–57.
Anderson, I. 1989. A mouse with a human immune system. *New Scientist* 14 January, 33–34.
von Boehmer, H. and Kisielow 1991. How the immune system learns about self. *Sci. Am.* **265** (4): 50–59.
Engelhard, V. H. 1994. How cells process antigens. *Sci. Am.*. **271** (2): 44–51.
Grey, H. M. , Sette A. and Buus S. 1989. How T cells see antigen. *Sci. Am.* **261** (5): 38–46.
Laurence, J. 1985. The immune system in AIDS. *Sci. Am.* **253** (6): 70 – 81.
Marrack, P. and Kappler, J. 1986. The T cell and its receptor. *Sci. Am.* **254** (2): 28–37.
Newell, J. 1990. Enzymes a la carte. *New Scientist* 24 March, 24–27.
Scientific American Magazine. 1993. Special Issue, Life, Death and the Immune System. *Sci. Am.* **269** (3): 1–124.
Tonegawa, S. 1985. The molecules of the immune system. *Sci. Am.* **253** (4): 104–113.

8. Science as Ideology

* Anita Rattan and ** Suresh I.S. Rattan

* *Feminist Researcher, Musvaagevej 6, 2tv, DK-8210 Aarhus-V, Denmark.*
** *Biologist, Laboratory of Cellular Ageing, Department of Molecular and Structural Biology, Aarhus University, DK-8000 Aarhus-C, Denmark.*

The "pure" universe of even the "purest" science is a social field like any other, with its distribution of power and its monopolies, its struggles and strategies, interests and profits, but it is a field in which all these invariants take specific forms. *Pierre Bourdieu.*

Science operates within the constraints of social conditions which determine how and what scientific knowledge can be produced. At the same time these social determinants derive their validity and legitimacy from the principles set forth by the scientific knowledge itself. The ideology of the ruling class naturalises existing social relations and represents them as eternal and unquestioned social facts by construing itself as a part of the field of scientific discourse. Science and ideology are thus conflated in the sense that the structures of institutional and non-institutional power are reproduced by recreating the logic inherent in the scientific method itself.

Science and technology have come under heavy attack from a wide variety of perspectives. The starting-point of such critiques has been to raise the question of ethics in abstract terms of the "use" and "abuse" of science in its applications. For example, in biology, ethical questions regarding genetic engineering, gene therapy, genetic screening, agricultural genetics and deliberate release of genetically engineered organisms into the environment have been collectively termed the issues of "genethics". Similarly, some other critiques have criticised biological determinism following the use-abuse model in terms of the appeal to biological principles for reinforcing social differences related to gender, class and racial categories.

Such criticisms have generally come from social scientists involved in analysing the effects of science and technology on different societies, from environmentalists who warn of the dangers of ecological change and damage, from health activists criticising the over-production and misuse of medicinal products, from animal-rights groups resisting the use of animals in biological research, from liberal political groups, and from religious fundamentalists who view science in opposition to their systems of belief.

However, other critiques of science have also been produced which have questioned the authoritative status of science as embodied in scientific objectivity, rationality and the notions of value-free science. Feminists and other radicals have been the main proponents of such critiques, and have revealed various kinds of distortion, bias, prejudice and assumptions in the scientific method and in the interpretation of its results. The main focus of these critiques is the so-called "scientific understanding" of the natural world, and the mechanistic and reductionistic philosophy on which this understanding is often based.

In this article, we aim to discuss the social character of science, its relationship to the growth of capitalism, and the ways in which certain scientific discourse constitutes the basis for our understanding of the social and the natural world. In addition, we will briefly discuss some of the limitations of reductionistic science and point towards the future trends in biology that may be relevant to its social character.

POLITICAL ECONOMY OF SCIENCE

Since modern science itself is a product of capitalism, it is subjected to the same rules of the game that govern the capitalistic relations of production. Therefore, like any other commodity in the so-called free market, science and technology are accessible to anyone who can pay to buy them. Science itself has acquired a use and an exchange value. The upsurge of multinational biotechnology companies and pharmaceutical and petrochemical industries in recent times is a mere logical extension of this characteristic of modern science as a profit-making business.

Obviously, this commoditisation of science is intimately linked to the tremendous success of biology, biotechnology and medicine in improving the life and health of human beings. Since, in principle, the products of science have universal applicability, science as a product itself has become a high-demand commodity in the international market. Therefore, it is not surprising that the circulation, distribution and availability of science, including technology and scientific literature, are also governed by the rules and regulations laid down by the politics of the international market, and are related directly to the purchasing powers of international capital. In other words, production and distribution of modern scientific knowledge is governed by those very social conditions which have come into existence since the rise of science and technology.

During almost two decades of Reagenomics and Thatcherism in the West, public money available for research was reduced progressively. At the same time, however, the number of private companies investing money in biomedical and biotechnological research has increased several-fold. An estimated number of several thousand large and small commercial organisations in the USA, Europe and Japan have made investments of the order of several billion dollars in biotechnological research and development. Obviously, the aim of such investments is not to generate "pure" knowledge. Instead, as Stanley Adams, the famous whistleblower of a giant pharmaceutical company Hoffmann-La Roche, once wrote, "Drug companies do not exist for the good of mankind, they exist to make profits".

Although, as a result of private funding, the total money available for biological research has increased significantly, its distribution is controlled according to research priorities set from the view-point of profit-making financial organisations and their political ideology. Obviously, those research projects get highest priority which may have the maximum commercial viability (or political significance) in the shortest time possible.

This is not to say that there exists a non-commercialised field of "pure" scientific knowledge which carries some intrinsic higher symbolic value. There is no real opposition between commercial and basic research within the existing framework of science and its practice. What may be termed basic today can generate several commercial applications under specific social and historical circumstances. Similarly, commercially oriented projects in biotechnology can generate entirely new information about the fundamental processes in biology.

Therefore, we view this opposition between commercial and non-commercial research as an unreal distinction. This is because, in principle, the scientific method to achieve results (having an academic or a commercial value) is the same in both cases. Apparent differences between them are due to the differences in the conditions of their production, such as research in an academic institution versus company-based research. Even these differences are disappearing rapidly because contract research approved and financed by the companies is being done increasingly within academic institutes, and also because of the creation of the so-called "science parks" where universities and industries work together under one roof. Thus, what has actually happened is merely that, with the flourishing of biotechnological companies, the problem areas have been redefined within the established framework of scientific method. This shift in itself is a logical necessity and an inevitable consequence of the ways in which capitalism expands.

The commercialization of scientific research and its expansion under new international economic order follows the same pattern as the world-wide growth of other commercial organisations. Because of the successful campaigns by the trade union and labour movements in increasing salaries and other benefits for the work force, the rising costs of research and development have made it necessary for biotechnological companies to look for cheap scientific labour elsewhere, in the same way as any other big industry. Thus a kind of intellectual division of labour is being created, in which original discoveries and inventions remain in the hands of the developed countries, whereas routine application, testing and development is being done in developing countries. Recent interest of several biotechnological and pharmaceutical companies in establishing new laboratories in many developing countries is an example of this neo-colonisation.

Furthermore, scientific discoveries and the resulting products are in circulation in the world within the constraints imposed by the policies of international capital. This is because the ways in which this scientific knowledge is being produced, marketed and protected by patent laws make its availability in the Third World only in terms of its monetary and political gains. In this way, scientific knowledge and technology constitute also a kind of symbolic power. To the extent that major research projects, such as the giant project for sequencing the total human genome, become a symbol of political supremacy.

THE SOCIAL AND POLITICAL POWER OF SCIENCE

Scientific knowledge is produced within social institutions. A whole infrastructure of the state is active in formulating national science policy, creating and managing grant-giving bodies, and rationalising productive structures, and thus in determining the planning and the direction of scientific research. Political priorities set for scientific research, the prestige of certain research areas in terms of monetary and symbolic profits in the national and international markets, local and general competition, the situation of research programmes at the centers or on the peripheries of national and international capital, and the allocation of resources regulate and monitor the production, distribution and circulation of scientific knowledge. These determinants constitute what may be termed as the external aspect of the social character of science.

There is, however, another aspect which is internal to science and concerns itself with the object of investigation and various possible modes of enquiry to arrive at those objectives. Certain positivistic ways of thinking claim that the "scientific method" ensures its objective and neutral character. This implies that at any given time, and in any given field of science, the choice of certain ideas to be tested and others to be rejected is determined purely by some inherent truth value of the idea. However, the real conditions under which the scientific method is applied show that, in a field of competing ideas, defining the problem, selecting the method for its solution, and interpreting the results are determined by the background knowledge accumulated through similar procedures, by the degree of testability, by the availability of techniques, and by the previous theoretical base. Therefore, it is impossible to separate the social character of science from the internal logic that constitutes the scientific method and that is also built into the productive infrastructure of science.

The production of scientific knowledge is a social field and is governed by the same laws of the market that control the production of any other commodity. Thus, scientific knowledge is a commodity as it is entailed by social capital in terms of the state infrastructure, the present and the future values of ideas and their recognition in the scientific field. In terms of scientific expertise and competence, it represents symbolic power appropriated by the interests of the dominant class.

Although science via technology intervenes actively in the processes of production, its role in transforming the social relations of production has been successfully converted by the interests of capital into one of subversion. This is not in accordance with the predictions of Karl Marx that the expansion of science and technology, a natural outcome of capitalistic growth, will create contradictions in the forces of production and class structure followed by the ultimate overthrow of the bourgeois order. This is because bourgeois ideology in reproducing and legitimising the conditions of production and reproduction of capitalism, of which science is also one of its products, refers to the same mode of reasoning which constitutes the basis of scientific knowledge. Science and ideology are thus conflated.

Science, therefore, has become authoritative. The objective positioning of science is relatively autonomous, and the status of indeterminacy, that nevertheless is conferred upon it by the interests of capital, has situated science uniquely in terms of the articulations of power relations. It implies that meaning becomes self-referential within the discursive field of science, and that the laws governing the truth of scientific knowledge refer back to their own bases.

In other words, the rules of the scientific objectivity, neutrality, and the certainty of scientific knowledge derive their validation from the mode of reasoning in science itself. Science has become a legitimising agent which, through the ideological state apparatus, neutralises the contradictory forces in social relations. The institutionalized form of power, as manifested in economic and political structures, legitimises itself by referring to a "true and certain" knowledge that emerges from scientific discourse. And the pretensions of science to revealing "truth" are rooted in scientific method, supposedly free of social and historical determinants, such that any kind of methodological criticism is carried out within its own premises.

Through a historical analysis of the progress of science, it has been shown that the inherent neutrality of science developed from the social and political forces of the seventeenth century, and is not the product of some internal logic of modern science. The separation of science from the social and the political aspects was a historical necessity for the institutionalization of science during that period. The rise of science as a cognitive programme and the rise of science as a social structure are connected in such a way that the dominant modes of cognition have been institutionalized in accordance with the so-called requirements of human progress, enlightenment and emancipation.

THE FEMINIST QUESTION IN SCIENCE

In the early period of the women's movement, feminists have explored and scrutinised male bias in various disciplines such as sociology, anthropology, history, psychology, media studies and literature. However, in recent times feminists have begun to look into the field of natural sciences for male prejudice operating there, and for possible solutions to this.

Since the prime object of feminist theory is to understand the position of women in society, feminists have analysed sex relations in terms of how they are organised in a society, how they constitute our experience, and how they become the organising principle of social relations in which they are produced. As feminist theory is not a monolithic body of knowledge, feminist critiques of scientific discourse, of scientific truth, and of implications of science to social phenomena arise out of several modes of thinking within feminism, which are reflected in the diversity of political practice.

One of the central questions posed by feminists has been to enquire whether it is possible to transform science in such a way as to create an alternative or feminist science. Some feminists reject the extension of biological theories, such as those explaining evolution and the functioning of the human brain, into the construction of theories in the field of social science. They argue that the claims of sociobiology and social Darwinism that extend to the economic and political orders of societies reduce humans from their status as a social being to that of a mere biological entity or a gene machine. Social sciences then produce the categories of gender differences and the notions of masculinity and femininity grounded in their biological differences. The feminist rejection of this dominant trend is because science claiming to be objective and neutral imparts its authority and power to theories in the social sciences that construe existing gender divisions as biological in origin and therefore natural.

Taking biological determinism as the initial point of debate, feminists have moved on to another terrain of reconceptualizing the broader relationships between science and society. For example, it has been argued that a certain scientific discourse is constituted by western thinking in its tendency to understand natural and the social phenomena in terms of dual opposites, such as nature-culture, male-female, rational-emotional, subject-object, and activity-passivity, which are asymmetric. The values thus associated with men are also associated with science, and the object of knowledge, Nature, is associated with femaleness. By implication, male values are considered to be higher in status than the female values.

Therefore, discussions in feminism and science have conceptualized the possibilities of constructing another science which is not grounded in these hierarchic dichotomies, and which does not see any rigid boundaries between the subject of knowledge and the natural object of knowledge. Although such critiques are relevant and necessary, the category of nature in such eco-feminism is constructed employing metaphors that operate at the same plane of discourse as that which constitute certain rational discursive practice.

Within feminism, technology for the regulation of reproduction has been the object of extensive debate because of its social and cultural meanings concerning women's relationship to motherhood, technology, and reproductive choice. Modern techniques, such as fertilization *in vitro*, fetal transplantation, fetal diagnosis and others have created a complexity of issues regarding the control of women's bodies by technology, and its implications in certain sociality of science. For example, motherhood has a double meaning in biological discourse, in the sense that biology constructs a notion of biological motherhood conflated with the bodily and social experience of "being" a mother. Therefore, feminists have raised questions of women's relationships to biology and technology.

Other feminist critiques of science as social practice can be classified into various groups, corresponding to the political understanding about the women's situation. Some feminists, while accepting that there are male prejudices in science, do not see anything wrong with science itself. That is to say that the basic methodology within which science is practiced is not a problem for them. They argue that the domination of male ideology in science is due to the fact that women are under-represented in scientific practice. According to their views, a possible solution of this would be to recruit more women in science.

A more radical criticism takes the above critique further, by arguing that the predominance of men in science has led men to a bias in choosing and defining problems, particularly in the areas of health sciences. It is claimed, for example, that contraception has been given importance only from the point of view of men primarily focusing on the techniques to be used by women. Similarly, it is pointed out that very little attention has been paid to the problems associated with menstruation and other issues of concern to women.

It is obvious from the conceptual understanding of such critiques that they fall under the political terrain of bourgeois liberal ideology. Since it is within the fundamental premises on which scientific practice functions that solutions to women's problems are to be sought in terms of equal opportunity, feminist voices on this side of the political spectrum, though indicating a necessary step towards equality, remain unproductive in conceptual terms.

Therefore, in order to conceptualise women's relationship to science, feminists need to employ the tools taken from psychoanalysis, in the sense that how a certain rational discourse in the imaginary order is alienating woman's body from its desire. A feministic critique of science developed in the space between the institutional practice of science and social construction of its concepts will be a precondition for the use of knowledge as a tool of liberation rather than of domination.

REDUCTIONISM AND THE LIMITATIONS OF BIOLOGY

The claims about the certainty of scientific knowledge are based on the principles of reductionistic scientific methodology, originally expounded mainly by philosophers Francis Bacon and Rene Descartes. Whereas Bacon changed the meaning of the nature and the goal of scientific enquiry such that empirical investigation was the only valid method of attaining knowledge, it is Descartes who is regarded as the founder of modern philosophy, because he raised skepticism in all forms of traditional knowledge by questioning the underlying method of arriving at an understanding of the natural world.

The mechanistic view of the world is based on a complete division of matter and mind where the universe of matter is devoid of any purpose of its own and can be understood fully in terms of mathematical principles. According to this view, human beings and all other living organisms function exactly in the same way as do machines, and the nature of their structure and function can be understood by applying the same rules that operate in machines. It implies that phenomena in the natural world can be reduced to certain eternal laws, and in order to formulate these laws, the phenomena can be broken down by the analytical method into smaller parts which can then be reconstructed in a logical way in order to gain an understanding of the whole system.

Although reductionism as a methodological principle has led to the tremendous success in the development of science and technology, and its intervention in transforming social relations in bourgeois society, reductionism as an epistemological principle that implies a claim to an absolute and certain knowledge is a false claimant. This is because, if nature can be reduced to mathematical descriptions by the analytical method, a correspondence between the external world as it is and its description in terms of the scientific facts is assumed. It is also assumed that there is a relation between the knower and the object of knowledge as unproblematic in the sense that the subject (knower) is outside this transparence of the object and its description. The knowledge thus produced is based on the principles of quantity, identity and non-contradiction rather than quality and contradiction, which lead to the criteria of objectivity in science and grant a status of ultimate authority to it.

Whatever the successes of reductionism as a method to obtain knowledge about the natural world may have been in the case of physics and chemistry, it has been almost a failure in contributing to our understanding of living organisms. For example, in biology, although mountains of information are available about the physical and chemical characteristics of DNA, RNA, proteins and other macromolecules, we are nowhere near solving the functional and organizational relationships among different components of the "living machine". The relationship between the genotype (the total genetic information of an organism) and the phenotype (the final physical outcome of that information, for example the shape of the nose) is a complete mystery.

Although reductionistic biochemistry has established similarities at the level of biochemical molecules and biochemical principles among all life forms, extreme differences encountered in the biology of different organisms in terms of their life histories, growth and reproductive patterns, and lifespans are least understood. For example, in the case of ageing and longevity, there is no explanation why two organisms belonging to different species should have different rates of ageing and maximum longevities when there are no differences between their fundamental biochemical processes. Similarly, although within the same species, say humans, the genetic and biochemical similarities between any two individuals may be more than 99%, it is obvious that their biological (not considering, social) differences are much more than that. Reductionism has been a total failure as a method to understand biology, irrespective of whether one considers an apparently simple phenomenon of cell division or more complicated phenomena such as the working of the human brain, memory organization, the functioning of intelligence and creativity, or the limits on lifespan.

Yet it is the philosophy of reductionism which forms the backbone of our present-day notions of medical treatment, gene therapy, genetic engineering, reproductive technologies, and agricultural genetics. After the initial euphoria of "one gene, one effect" was over and it was realized that single gene manipulations are not the most efficient or the most appropriate way of dealing with biological phenomena, holistic ideas in biology have been developing in order to understand biological, social and environmental issues. Even at the level of genes, the concepts of split genes, shared genes and virtual genes are a long way from the principles of reductionism.

Biology is now entering a phase of deconstruction, in which radical changes in our ways of looking at biological problems are already occurring. The social accountability of scientific research is taking a central stage. The myths of pure knowledge and of the eternal search for truth are being replaced by discreet material realities of relevance, ecological balance, and human equalities in scientific research. Perspectives in biology are changing and a new dimension in the web of life is opening up.

9. Biodiversity and Human Welfare

N.H. Ravindranath

Centre for ASTRA and Ecological Sciences, Indian Institute of Science, Bangalore 560 012.

Biological diversity which is simply biodiversity (BD) for most is currently top on the agenda of governments, UN agencies, industries, NGOs, researchers and local communities. There is an unprecedented surge of awareness and concern on the impending threat to biodiversity and the need for its conservation. Declines of biodiversity is one of the top global environmental problems along with climate change, tropical deforestation, ozone depletion, acid rain and land degradation. In fact these global problems are linked to one another; for example the adverse impact of deforestation and climate change on biodiversity. The unprecedented global concern on biodiversity stems from its actual and potential use or value to humans and the potential threat to it. The real threat of extinction of popular animals such as tigers, lions, African elephants and pandas has caught the imagination of all including children around the world. And most are aware of some organized and focused efforts to conserve these large mammals.

In the short period preceding the 1992 UN Conference on Environment and Development, and in its aftermath there was a flood of literature on various aspects of BD; Report on Biodiversity by World Conservation Monitoring Centre (WCMC 1992), The Global Biodiversity Assessment Report of United Nations Environment Programme (due for release in 1995), and Global Biodiversity Strategy (WRI-IUCN-UNEP, 1992) and similar reports by individual governments and other international organizations. Unlike other scientific problems, the threat to biodiversity and climate change requires global level assessment and concerted action, apart from action at the local and national level, which seems to be happening. In this chapter an attempt is made to present the issues related to the relevance of biodiversity to humans in a diverse world, threats to it, the need for and strategies for its conservation to promote human welfare in the long term.

WHAT IS BIODIVERSITY?

Biodiversity is commonly used to describe the number, variety and variability of living organisms. It encompasses the totality of genes, species and ecosystems. The terms are explained based on WRI-IUCN-UNEP (1992) and WCMC (1992).

Genetic diversity refers to variation of genes within species. It includes distinct populations of the same species (such as the thousands of rice varieties in India) or genetic variation within a population (such as different blood groups in humans).

Species diversity is one of the most commonly understood variations in living organisms. The number of species in a region or species richness is a common measure of diversity. But for a more precise measurement, "taxonomic diversity" considers the relationship of species to each other. The concept should also include the degree or extent of variation with which organisms differ widely from each other as it contributes more to overall diversity than those which are similar. The higher the difference of a species from others, the greater its contributions to any measure of global biodiversity.

Definition and Classification of Ecosystem Diversity

Ecosystem diversity is complex and difficult compared to genetic or species diversity because "boundaries" of communities or association of species and ecosystems are elusive. Ecosystem diversity also includes the impact of abiotic components such as soil and climate. Ecosystem diversity is often evaluated through measures of the diversity of component species involving an assessment of the relative abundance of different species as well as the consideration of the types of species (such as size of species, trophic level or taxonomic group).

Human cultural diversity could also be considered as part of BD. Cultural diversities are manifested by diversity in language, religious beliefs, land-management practices, social structure, crop selection and other attributes of human society.

One of the first steps in any discussion on BD is an inventory of species; which species, how many and where. The estimates of total number of species on earth at present vary from 5 to 100 million. Of this only about 1.7 million have been described to date. However, working estimates of BD suggest that there could be 12.5 million species on earth. The broad distribution of major groups of organisms is given in Table 1. When the working estimates are taken, the earth seems to belong to insects and microorganisms in terms of the number of species. But among the described species next to insects, humans seem to have concentrated on plants as 250,000 plant species have been described so far (WCMC 1992).

Given the importance and urgency of information on diversity of different areas, habitats or ecosystems, significant efforts are being made to understand the species richness patterns very rapidly (WCMC 1992).

BIODIVERSITY AND HUMAN DEPENDENCE

Humans have depended on other life forms as a part of the food chain. Hunter gatherers wholly depended on meat from a range of animals and wild fruits, roots, tubers etc from plants. The evolution of crop plants began between 5,000 and 10,000 years ago. Humans have domesticated a few hundred of the 250,000 species of flowering plants, a few tens of 4000 species of mammals, around 10 of the 19,000 species of birds, about 5 out of millions of species of insects. Next to food products, the most important end use of biodiversity is its use for medicinal purposes. The World Health Organization has listed over 21,000 plant names that have reported medical uses around the world. The use of plants for their medicinal properties exists among the widest spectrum of human societies from the indigenous communities to modern industrial societies.

Table 1. Numbers of species in the groups of organisms likely to include in excess of 100,000 species (plus vertebrates)

	Described species	*Estimated species*	
		Highest Figure	*Working Figure*
Viruses	5,000	500,000	500,000
Bacteria	4,000	3,000,000	400,000
Fungi	70,000	1,500,000	1,000,000
Protozoans	40,000	100,000	200,000
Algae	40,000	10,000,000	200,000
Plants (Embryophytes)	250,000	500,000	300,000
Vertebrates	45,000	50,000	50,000
Nematodes	15,000	1,000,000	500,000
Molluscs	70,000	180,000	200,000
Crustaceans	40,000	150,000	150,000
Arachnids	75,000	1,000,000	750,000
Insects	950,000	100,000,000	8,000,000

Notes: The figure for described species (mostly given to the nearest 5,000) were arrived at by consulting relevant specialists as well as by critically reviewing the literature. The highest figure estimates for existing species, many of them frankly speculative, are the highest encountered during a survey of recent literature. The 'working figure' estimates are conservative. The figure for bacteria has been arbitrarily capped at 100 undescribed to 1 described species on the grounds that projections involving more than two orders of magnitude are inherently unsafe. The biggest question marks lie over the true numbers of species of viruses, bacteria and algae. Substantial upward revisions from the working figures for these groups may prove justified with time. The figures for fungi, protozoans and nematodes are also insecurely based. Note that the Fungi and Protozoa are used in the 'traditional' sense, while Bacteria includes cyanobacteria. The figures for 'insects' include all hexapods, and that for described insect species assumes totals of 400,000 for Coleoptera, 150,000 for Lepidoptera, 130,000 for Hymenoptera and 120,000 for Diptera.

Reference: 1. WCMC (1992).

The dependence on biodiversity is highly diverse for different societies. Human dependence on diversity has varied with time as for example the number of varieties of rice cultivated in India has rapidly declined from tens of thousands to a few hundred with the advent of green revolution. Similarly the number of medicinal plants used is gradually declining with the advent of western allopathic medicines in developing countries. Given the diversity of climatic factors, population density, extent of modernization of agriculture, access to modern energy and industrialization, the human dependence on diversity is varied. In the following section the extent of dependence on BD and its impact on quality of life is discussed. Even as we move towards the twenty-first century, the dependence on biodiversity continues, though the extent of dependence continues to vary among different communities.

Indigenous communities, traditional agricultural societies, modernizing agricultural societies and modern industrial agricultural societies are the broad categories considered to understand the varying relevance of biodiversity.

INDIGENOUS OR HUNTER-GATHERER SOCIETIES

Indigenous or tribal societies are predominantly present in the tropical countries of Asia, Africa and Latin America. Agriculture in these societies is either absent or is very primitive. Humans in such societies depend wholly on naturally occurring plant products and animal meat within their catchment areas. They depend on naturally occurring BD for all their nutrition, fuel, fibre, medicinal uses, housing construction materials, implements and so on. In addition to meeting these basic subsistence needs, communities also trade in a range of plant parts as raw materials for processing elsewhere and also as finished handicrafts. In Central and West African countries more than 1500 species of wild plants – fruits, nuts, seeds, roots, teas, herbs and vegetables are included in their diets. In one region of Peru, fruits of 193 species are regularly consumed and 120 of these are exclusively gathered from the wild (WCMC 1992). A study of eight village communities in Ghana showed that 49% to 87% of their earnings came from the sale of such non-timber forest products. A similar study in Cameroon showed that subsistence gathering, fishing, trapping and hunting contributed to more than half of the local incomes. In North-West Amazonia alone, some 2000 species are used for medicinal purposes (Ryan 1992). An ethnobiological study conducted in an Indian peasant market of Patzcuaro in Central Mexico revealed a total of 222 products corresponding to 138 species of plants, animals and mushrooms (Table 2). Among the products used, foods and medicines dominated with respect to number of species. It is also interesting to note the dependance on a large diversity of natural and modified ecosystems (Table 2). For example foods were obtained from lakes, forests, grass land, shrub land, kitchen gardens, apart from cultivated fields. Thus, traditional societies depend on a large diversity of plant and animal species and products from a diversity of ecosystems in their surroundings for food, medicines and livelihood.

TRADITIONAL AGRICULTURAL SOCIETIES

Traditional agricultural societies have spread all over the tropical world, particularly in Sub Saharan Africa, and to some extent in South and South-East Asia and Central and South America. Though these societies depend partly on naturally available food sources, their food requirements largely come from crops and domestic animals. However, for many of their requirements they depend on naturally occurring plant products. A study of a traditional agricultural community in West Bengal in India showed that 155 species were used by the communities. Of this, 73 plant and 2 animal species were used regularly. In all, 11 different parts of the plants were used for various purposes such as fuel, fodder, medicine, commercial, household articles, religious, ornamental and recreational purposes (Malhotra *et al.* 1990).

Table 2. Number of Plant, Mushroom and Animal Species by use Category and Ecosystem Type in Indigenous Markets of Pa'tzcuaro (Mexico)

	Natural Ecosystems				*Modified Ecosystems*			
	Lake	*Forests*	*Shrub-lands*	*Grass-lands*	*Cultivated fields*	*Fallow fields*	*Kitchen Gardens*	*Total*
Foods	19	9	6	5	15	9	26	89
Medicines	1	13	3	0	0	1	13	31
Ornamentals	0	2	0	0	1	0	2	5
Forages	0	0	0	0	2	0	0	2
Fuelwoods	0	4	0	0	0	0	0	4
Domestics	2	1	0	0	0	0	0	3
Work/Tools	0	4	0	0	0	0	0	4
TOTAL	22	33	9	5	18	10	41	138

Reference: Maps, C. and Toledo, V.M. (1991) The indigenous market of Patzcuaro Mexico: An ethnobiological study, *Journal of Ethanobiology*. Mapes and Toledo (1991).

The traditional agriculture is characterized by the absence of any fossil fuel use and dependence on a large diversity of local varieties of crops and breeds of livestock. Farmers also adopt a variety of mixed species cropping. Farmers in the Gonder areas of North-West Ethiopia are reported to plant more than 6 crops together in their backyard; maize, faba bean, sorghum, cabbage, tomato, potato, pumpkin and bottle gourd. This also acts as an insurance against crop failures. Apart from dependence on a diversity of crop species, a large diversity of cultivars are grown by traditional farmers. A detailed study in the Nuba Mountain area of Sudan has shown the creation and maintenance of traditional cultivar diversity of sesame (Bedigian 1991). In the Nuba region, dozens of traditional sesame varieties are present. Farmers growing sesame in small plots grow a mixture of cultivars, which are adapted to the region but which differ from one another in date of maturity, seed colour and response to prevailing diseases and pests. The primary goal is to obtain reliable yields each season for local consumption. The cultural and ecological diversity of traditional farming systems in ancient agricultural regions has shaped genetic variation within traditional crop varieties. Thus traditional agricultural societies, like indigenous societies, depend on the large BD for survival.

MODERNIZING AGRICULTURAL SOCIETIES

Large parts of South and South-East Asia fall into this category where farmers are moving towards agriculture based on specialized crops, HYV (high yielding varieties) and fossil fuel use. In India, for example, 85% of area under wheat (20 Mha) and 66% of area under rice (28 Mha) was under HYV during 1990 (TSL 1992). Local varieties of millets, pulses, oil seeds and cotton etc. have also been slowly replaced by HYV; by exotic as well as local HYVs. There is a change in diversity of crop species as well as cultivars or varieties with expansion of trade in food grains within and across nations. Specialization in crops is also increasing with regions dominated by crops such as sugarcane, rice, cotton and wheat dominant districts in India.

Even afforestation/reforestation programmes in modernizing agricultural regions like in the North-West India (Punjab and Haryana) or Western Ghats regions, or semi-arid tracts of South-India are dominated by a few exotics such as *Eucalyptus spp, Acacia auriculiformis, Tectona grandis* and *Pinus spp.* (Ravindranath and Hall 1995). This is true of industrial plantations all over the tropics (FAO 1993). Farmers also had the tradition of growing a large diversity of trees on farms, along avenues, near homes and in temples. Homestead gardens in Kerala, Western Ghat region of South India, Sri Lanka and Indonesia are known to have a large diversity of tree species (Ravindranath and Hall 1995). A semi-arid village ecosystem, in South India, of 330 ha with 240 ha under cropping and a population of 932, had 47 tree species. The same village where over 50% cropped area was under HYV of rice and where farmers use fossil fuels, a study found that components from over 50 plant species were used for over 30 end uses apart from use as food (Ravindranath and Somashekar 1991).

INDUSTRIAL SOCIETIES

With selection and breeding, HYVs have been developed and adopted in industrial agriculture. Large monoculture and monospecific cropping and tree plantations are generally features of these societies with high fossil fuel inputs and high yields. The highly developed transportation and trade along with high purchasing power, enables them to specialize in selected crops to achieve high efficiency of production and lower costs. *In situ* biodiversity is of little relevance or consequence in these societies as all seeds used in cultivation come from commercial seed company sources. The foods used are often imported from different continents and one can find fruits and vegetables from both tropical and temperate regions sold in all food stores in Western Europe and USA.

Why biodiversity is a global issue today

Humans have always depended on BD over centuries. But till recently it was never a global level issue. Possibly there are two reasons for this, where one is a result of the other. Firstly, there has been an unprecedented decline and loss of BD due to anthropogenic causes. Secondly, there is a growing realization of the immediate loss of sources of nutrition, medicine and livelihood in indigenous and agricultural societies and of large future potential or relevance of BD to food and medicinal value in industrial societies.

The causes for decline and loss of BD have been generally well documented. Some of the dominant causes are i) loss and fragmentation of habitats, ii) overuse of natural resources particularly plant biomass, iii) introduction of exotic species, monocultures and genetic uniformity in agriculture and forestry, iv) air, water and soil pollution, iv) loss of control over communal land and natural resources to indigenous and local agricultural communities [WCMC (1992), UNEP (1995) and WRI-IUCN-UNEP (1992)]. Regarding animal species extinction, a world conservation union analysis showed that since 1600, 39% of the extinction resulted from species introduction, 36% from habitat destruction and 23% from hunting and deliberate extermination (Fig. 1).

Habitat loss is the root cause of the large decline in BD. One of the most important and glaring habitat losses is conversion of tropical forests to non-forest uses such as cropland, pasture land, infrastructure and simply often to be left barren. Tropical forests are estimated to be the home for 40%–90% of world's BD (WRI 1994). Humans have been clearing forests for thousands of years probably at a lower rate. Recent systematic monitoring has shown that during the decade 1981–90 the annual rate of deforestation in tropics was 15.4 Mha (FAO 1993). If the same rate is assumed for the past 25 years, the total area deforested since 1970 is estimated to be 385 Mha. This is larger than the size of Indian sub-continent which is the home of over a billion of human population in 1990. Fragmentation of forests (FAO 1993) and forest degradation due to fire, livestock grazing and fuelwood extraction also contributes to loss of plant diversity. According to a review by Myers (1993), we are losing 27,000 species per year in the tropics alone. This rate is 120,000 times higher

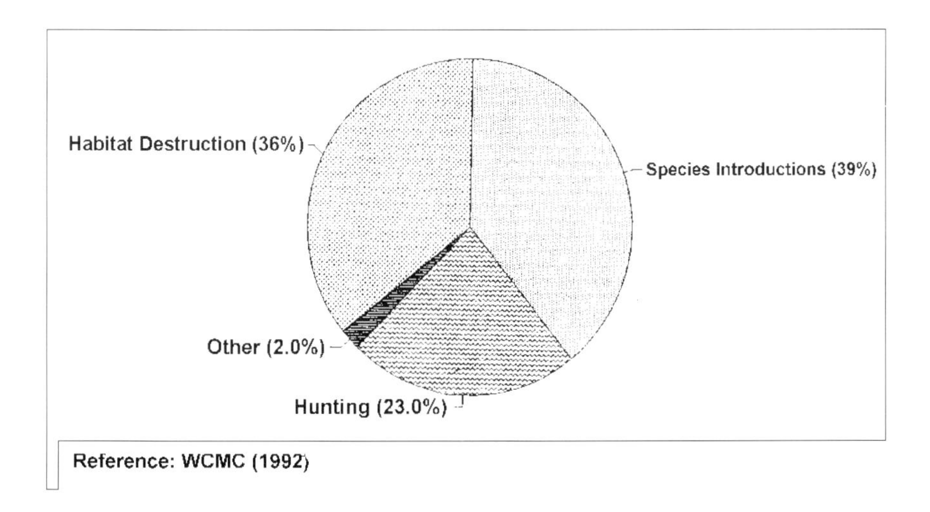

Figure 1. Known causes of Animal Extinction since 1600
Reference WCMC (1992)

than the rate before the advent of human era. Further, the world faces the prospect of losing 20% of all species within the next 30 years and 50% or more by the end of next century.

Thus deforestation, forest fragmentation and forest degradation are the dominant causes for loss of BD. These also lead to soil erosion, disruption of watersheds, floods, shortage of water and loss of productive capacity of soil further indirectly contributing to decline in forest regeneration leading to loss of BD.

Even wetlands are being lost. The United States has lost 54% of its wetlands, New Zealand has lost over 90% and many parts of Europe have lost nearly all their wetlands (Dugan 1990). Tropical countries like Chad, Cameroon, Niger, Bangladesh, India, Thailand and Vietnam have lost over 80% of their fresh wetlands.

As against a dependence on a large diverse genetic base and mixed cropping as observed in Africa and India in traditional agriculture, today the genetic base of crop production is very narrow. As can be observed from Table 3, both in modernizing and industrialized agricultural societies, there is a clear trend towards genetic uniformity of the major crops such as rice, wheat, potato and cotton. For example, in India over 30,000 local varieties of rice were in cultivation traditionally but by 1990, 66% of the rice area is under a few HYVs and 75% of production comes from less than 10 varieties. As a result of the green revolution strategy of introduction of HYVs and modern inputs, crop production has registered dramatic increase in Asia. But in the process, all the native diversity of crop varieties has been lost. This is true of even other crops grown by the poor farmers in the tropics such as maize, sorghum, cotton, and oilseeds.

The most disturbing fact is that with the continued growth in human and domesticated livestock populations, the rate of loss of habitats is projected to increase. For example, the deforestation rates are projected to increase to 20 to 23 Mha per year by year 2025 and 73% of all tropical forests are expected to be cleared by 2100 (IPCC 1992). The cropland area requirement is projected to increase along with population growth. Such an increase might require 50% more land under crops. The demand for industrial wood and fuelwood in the tropics is estimated to treble by 2050. Further, with modernization of agriculture and the introduction of HYV the large diversity of local cultivars are being permanently lost from human settlements in the developing countries. Projections show that between 1% to 11% of the world's species are committed to extinction between 1975 and 2015 (Reid, 1992). The projected climate change could further complicate as climate is expected to change faster than the species ability to adapt.

Though massive losses of BD are inevitable, current knowledge on the rate of species loss, BD depletion processes, evolutionary consequences, what efforts are required, and how soon, is very inadequate (Myers, 1993).

Table 3. Extent of genetic uniformity in selected crops

Crop	*Country*	*Number of Varieties*
Rice	Sri Lanka	From 2,000 varieties in 1959 to 5 major varieties today 75% of varieties descended from one maternal parent
Rice	India	From 30,000 varoetoes tp 75% of production from less than 10 varieties
Rice	Bangladesh	62% of varieties descended from one maternal parent
Rice	Indonesia	74% of varieties descended from one maternal parent
Wheat	USA	50% of crop in 9 varieties
Potato	USA	75% of crop in 4 varieties
Cotton	USA	50% of crop in 3 varieties
Soybeans	USA	50% of crop in 6 varieties

Reference: 1. WCMC (1992).

Perception of indigenous and agricultural societies:

The immediate impact of the loss of habitats and decline in BD will be felt by the indigenous and agricultural societies. They will lose their source of nutrition, medicines and employment. These communities will also have to face the consequences of loss of top soil and its productive potential, declines in availability of soil moisture and climate change. Thus the indigenous and agricultural societies will not only suffer immediately but also in the long run due to land degradation. Surely they are concerned but their concerns may not find adequate expression in international mass media or scientific literature.

Perceptions of industrial societies:

The industrial societies (Western Europe and North America) facing large food surpluses (grains and dairy products) and who currently largely depend on synthetic formulations for medicinal uses are unlikely to be affected in the short run. Further, historically plant breeders, biotechnology and pharmaceutical companies have had unrestricted access to the developing world's genetic diversity. This situation is unlikely to continue in the post-Biodiversity Convention era.

The World Health Organization has listed over 21,000 plant names that have reported medicinal uses around the world. Only about 5000 higher plants have been thoroughly investigated as potential sources of drugs. The US National Cancer Institute has identified over 1400 tropical forest plants with the potential to fight cancer. One such example is the Rosy Periwinkle *Catharanthus roseus* native to Madagascar. This plant used for generations by tribal healers, is now used in the production of drugs, effective against Hodgkins disease and other forms of cancer.

According to a study for 1985, the total value of plant based drugs in US alone was the US$ 18 billion (WCMC 1992). If taken at the global level, the value of plant based commercialized as well as traditional medicines would be very large (according to Myers (1993) it is about US$ 40 billion), indicating the economic value of plant diversity for medicinal purpose alone. In spite of the advances in medical science and biomedical engineering, there are many diseases for which no effective treatment is known. There are unknown and untried plant species which may have potential as new drugs to cure these diseases (WCMC 1992). In future some new diseases may appear for which cures could be possibly found in the plant diversity currently existing in the tropical forests.

Industrialized, industrializing and indigenous societies are all concerned about the decline and extinction of some of the known and understood plant species. In addition, there is a growing concern that the hundreds of thousands of species not yet analyzed for potential use in crop plant breeding and for therapeutic values may become extinct. Even in India where over 2500 plant species are used by traditional healers species; *Aconitum, Dioscorea and Ephedra* are some of the medicinal plants under threat in the wild. *Dioscorea deltoidea* used in the manufacture of contraceptive pills is on the decline due to over collection (WCMC 1992). Many large multinational pharmaceutical companies are also sensing large profits to be made from exploitation of BD. Even large seed companies and International Crop Research Institutes (such as IRRI) are concerned about the loss of genetic diversity such as the wild relatives of cultivated plants and varieties cultivated by traditional farmers (land races) for breeding for pest resistance, salt tolerance, differing durations, climate changes and ultimately high and sustainable yields. The economic importance can be observed from the fact that the top 25 agricultural biotechnology – or crop breeding-firms spent US$ 300 million on research and development in 1988 in the US alone (Hobbelink 1991). Crop breeding has generated large returns on investment to the companies, nations and farmers.

CONSERVATION OF BIODIVERSITY

Conservation of BD includes conservation of genes, species, populations, communities and ecosystems. Humans are present everywhere and they are a part of the ecosystems which are influencing the biological resources continuously. Unlike the earlier conservation approaches, which were based on narrow and often specific species or habitats, the biologists have started taking a holistic view and are considering non-biological factors in determining BD conservation measures. The need for conservation of BD evokes support from every section of humanity; from local communities to local governments to national governments to industry (local and multinational), mass media, the scientific communities and even children. Rarely has humanity shared such a common concern and is willing to work towards the goal of conserving a resource for local and global good. There is general agreement on the need for conservation and enhancement of BD, but there are serious disagreements on how to achieve that or the *modus operandi*. This is to be expected given the diversity of cultural, social and economic conditions of different societies, levels of benefits currently derived, differing perceptions of the future benefits and the costs involved.

Global concern reached a climax with the signing of the convention on BD by 158 Governments at the 1992 United Nations Conference on Environment and Development and later ratified by most governments. National governments, UN agencies (like UNEP, UNESCO), International NGOs (like IUCN, WWF), local NGOs, World Conservation Monitoring Centre (WCMC) and individual scientific groups and citizen groups have all developed or are in the process of developing strategies to conserve BD. Several good strategy reports are already available such as WCMC report on Global BD (1992), Global BD Strategy (WRI-IUCN-UNEP,1992) and Global BD Assessment (UNEP 1995, due for release), and in addition several country strategy reports are available detailing the action plans. WRI-IUCN-UNEP (1992) has listed 85 action plans in the 'Global Biodiversity Strategy' report. Thus the strategies do not merit repetition here instead some of the strategies are simply listed and only a few strategies are briefly discussed. The strategies include conservation of genetic diversity, species, habitats and ecosystems, and broader integrated approaches. The strategies include local, regional, national and global level actions.

i) *In situ* conservation of threatened plant and animal species and wild relatives of crops
ii) Ex situ conservation of plants, animals and microbial diversity
iii) Regulated trade in wildlife products
iv) Survey, identification and monitoring of BD
v) Sustainable modes of use of natural resources – techniques and institutions
vi) Wildlife protection acts
vii) Forest conservation acts
viii) Ecodevelopment of communities depending on BD
ix) Increasing agricultural productivity

One of the contentious issues is that the bulk of BD exists in the tropical countries which also happen to be the developing countries; 14 of the 18 global BD hot spots defined on the basis of high plant endemism and threats due to human activity are located in tropical countries (Myers 1990). Though conservation of BD is beneficial to all, some perceive that there will be costs in the short term particularly for the poor who depend on the sources of BD for livelihood. Even among tropical countries there is a gradient of development ranging from Malaysia and Thailand to some of the poorest countries in Sub-Saharan Africa. It is important to note that poor and particularly indigenous communities are on the frontline and are likely to be the first to suffer and suffer most both in the short and long term. Thus poorer communities have a larger stake in conservation compared to the rest in the developing world itself and in industrial countries. Indigenous and agricultural communities need to conserve BD for their own sake and in the process others who live away will also benefit.

The fundamental issue is that the maintenance of the naturally evolved BD and its conservation and promotion by indigenous and agricultural communities will require information, dedicated human effort, even sacrifice of short term gains and possibly involve financial costs, for example reduction of grazing pressure or reduced rates of extraction. The existing BD has been retained and protected till today largely due to the efforts of these two communities. Thus the basic premise of all conservation efforts must be to ensure long term benefits and stake to local communities. Two of the most important factors that need to be addressed are, need to empower local communities to manage land and all the natural resources including BD and the other is to increase agricultural productivity to prevent forest conversion and to reduce the pressure on remaining forests and pasture lands.

In situ conservation

Much of the crop genetic diversity (including the wild relatives of crops) is in the custody of the farmers who follow age-old farming and land use practices. The primary advantage of the *in situ* conservation of traditional agro-eco systems is that crop populations continue to be influenced by the evolutionary processes that created and maintained the genetic resources. Genetic diversity is maintained in traditional agro-eco systems by cultural intervention as well as natural selection. Heterogenous environments created within agro-ecosystems not only provide a variety of selection pressures, but they also provide some crop plants protection from environmental stresses. Such protection is not provided by modern agricultural practices.

Similarly, livestock breeds in developing countries have evolved under conditions of environmental stress. There is likely to be competition for resources between humans and livestock in the future. Thus livestock of the future will have to survive under severe shortages of feed and changing environmental stress. Thus the value and use of indigenous livestock breeds in developing countries under environmental stress are likely to be more relevant in future.

Protected area and nature reserves

The second aspect of *in situ* conservation is that of natural habitats with all the species present in them, for example; forests, grasslands, mangroves, wetlands etc. The protected area concept is one of the basic approaches adopted to conserve BD in its natural state. Protected areas are characterized by varying degrees of controls and restrictions on human activities. The popularity of the protected area approach can be gauged from the area dedicated to this option; globally about 6% of the world's land area (792 Mha) is under varying degrees of protection and management. In North and Central America, 12% of area is under protected area management (WRI 1994). Even in a developing country like India, 16% of forest area is designated as a protected area. One of the major problems encountered in many protected areas is the conflict between the interests of local communities residing in and around the protected area and the authorities. The restrictions on traditional rights and uses enjoyed by the local communities are not acceptable to local communities. This problem could be resolved by involving the local communities in all decisions and by ensuring sustainable modes of extraction and land and product tenures to the locals to ensure long term stake.

Conservation by indigenous communities

Conservation of BD by indigenous communities should necessarily mean sustainable use of biological resources. However, a number of questions arise when considering sustainable management; the pressures of growing human and livestock population, shrinking land resources, increased exposure to urban environment, material wealth and modern technologies.

Commercialization and external demand for indigenous resources (like rattan, nuts, medicinal herbs) often leads to non-sustainable extraction from eagerness for quick money, particularly if the local communities have no long term stake. There could be conflicts here between the need to conserve BD resources through conservation of traditional modes of resource use, and the need to generate incomes to create incentives and stake for local communities to reduce any non-sustainable rates or modes of extraction.

Today there are very few communities or locations which are not influenced by the state, the urban world or even the external world (from tourism and demand for products originating from say tropical forests). The modern schooling system and urban areas to which the children of indigenous communities are exposed bring in a new value system to communities. Indigenous belief systems are the basis of traditional management systems; if these beliefs change, the traditional management systems are thrown off balance (Clay, 1991). A number of indigenous peoples have developed complex systems of resource management that maintain a sustainable balance between harvest rates and yields of locations. These resource management systems are deeply embedded in their cultures (Clay, 1991). Gadgil (1991) concludes that in India, none of the population segments practice prudence today and ecosystem (local) people are the ones most likely to become motivated to practice prudence. Further, communities are likely to practice prudence if they perceive the resource base as finite and limited. The social groups that are likely to practice sustainable rates and modes of use are the local communities living in the vicinity and who have control over the resource.

Another justification for the primacy to be given to the indigenous communities in conservation efforts is due to the knowledge communities possess about the behavior of complex ecological systems and their evolution accrued over generations. The indigenous knowledge on conservation and enhancement of BD could be put to best use as an integrated system of knowledge, practice and beliefs. This can be best achieved by promoting community based resource management systems (Gadgil *et al.* 1993). There is also a need for systematically recording the indigenous knowledge on all aspects of BD in different regions. This may necessitate the preservation of cultures of indigenous societies. However, there could be conflicts often where indigenous communities may want to modernize. But what is important is that such communities must have freedom and power to decide their destiny and they should not be forced to abandon their habitats and cultures as often happens due to logging for exports by large companies or by submergence due to hydro-electric power projects, meant to generate power for faroff urban and industrial societies.

Local community empowerment and management

BD conservation in the tropics is often simply termed as a land use issue, since land is by far the most important asset, particularly for producing food, in developing countries. The governments or the state forest departments claim to control most forest resources in most countries. It is difficult to visualize how a few tens of thousands of forest department staff be it in India or Indonesia or Brazil or Zaire can control, conserve and manage millions of square kilometers of forests where tens of millions of population live in and around forests and are depending on them. Loss of traditional local control over land, other natural resources and their management to local communities is at the root of degradation and loss of BD. Similar conclusion could be drawn for fisheries. If access to an area is open to any user, the BD in the commons (lands or waters) is likely to be degraded more quickly (Gadgil et al. 1993). The governments of tropical countries need to enact legislation to transfer control over land to local communities with adequate power to manage and to control access to outgroups.

Ecodevelopment of communities

Conservation of BD cannot be considered in isolation, particularly in locations involving indigenous and agricultural communities. The human and cattle populations are increasing. For example, in Africa human population is growing at over 3% annually and livestock population is estimated to increase by a factor of three by 2050 to 1. 69 billion (Zuidema et al. 1994) grazing on dwindling forest and pasture land. The crop productivities are low, thus necessitating conversion of forests to open forests to grazing land and to cropland. Even grazing lands are being subjected to over grazing, soil compaction and soil erosion. Thus to reduce pressure on forests and pasture lands there is a need to increase crop and livestock productivities. Thus BD conservation must be linked to enhancing food production, sustainable use of natural resources and socio-economic development of communities, regions and nations. BD conservation is a development issue rather than a simple species preservation issue as BD richness and poverty often co-exist. The practical option is not to halt industrial development and intensification of agriculture; instead, both need to be continued but with an emphasis on efficient, sustainable and equitable resource use (Gadgil 1992).

REGULATED ACCESS AND ROYALTY TO BIOLOGICAL RESOURCES

Till recently the commercial and industrial users have been able to use and to profit from BD using their technological innovations and commercial channels with no benefits flowing back to the local communities who have created and protected the diversity (Ryan 1992). However, one of the key principles of Biodiversity Convention is the regulated access to biological resources. Accordingly, the national governments will have full control over any access to biological resources particularly for the external agencies. To reward the countries of origin, one of the options considered is that the sovereign states would be financially rewarded for permitting the use of BD. Assuming that this arrangement works and the national governments get adequately financially rewarded for the BD, they will be free to deploy the receipts depending on their priorities. Unfortunately what may happen is that the indigenous and agricultural communities which have maintained and promoted the BD may not get any direct benefit. For example, the national governments may use the royalty receipts say to build a highway elsewhere or undertake afforestation elsewhere or expand the government machinery ostensibly to enhance government's efforts on conservation of BD. Such an arrangement is detrimental to the indigenous communities as well as BD. The local communities may have no stake in promoting BD in such an arrangement leading to decline in BD. However, this is a politically sensitive issue and all governments would like to retain full control over the fate of funds generated. In the long term interests of conserving BD and promoting development of indigenous communities, the national governments must make an *a priori* commitment such that a dominant and fixed share of all royalty received for BD should be passed on to the specific locations and to the local communities. There should be legally binding institutional arrangements to ensure the flow of financial rewards to the real protectors of BD. Equally important is the fact that the local community should decide how the financial receipts are used and not the bureaucrats in the capital city.

CONCLUSION

Humans have always depended on BD for survival and prosperity. The human population is not homogenous and thus the nature and extent of dependence on BD is different for indigenous, agricultural and industrial societies. The current and future generations of humans are destined to witness an unprecedented rate of loss of BD. The decline and loss of BD is likely to affect all, but most immediately the indigenous societies who depend on BD for nutrition, medicines and livelihood followed by the agricultural societies. But there is a realization that loss of BD which has happened due to loss of habitats, over extraction, pollution, modernization of agriculture based on genetic uniformity will be detrimental to all, including industrial societies. There is a need for conservation and restoration of BD for immediate gain as well as for long term benefits-sustained pollution free agriculture, for pharmaceutical industry for promoting human health, livelihood to local communities, quality of life from indigenous communities to industrial societies and for aesthetic, educational and recreational benefits.

There is a growing realization and consensus on the need for conservation but perceptions vary among different societies on how to achieve that. Given that large BD is present in tropical or developing countries, particularly in tropical forests and traditional agro-eco-systems, there can be little disagreement firstly, on the need for empowering local communities and to create a long term stake for them in conserving BD through institutional arrangements, sustainable harvest modes, appropriate financial rewards for maintaining and promoting BD. Secondly, to undertake sustainable and equitable development in the indigenous and agricultural societies through increased crop and livestock production to meet the growing demand for land, food, fibre, fodder, fuelwood and timber. Appropriate technology, financial support and institutional structures at local, national and global level are required to promote prudent use of natural resources, sustainable and equitable development. Fortunately, every country is in the process of developing strategies to support the global efforts to conserve BD. But surely many of the actions required to conserve BD may not be planned and implemented as the clock is ticking and the time is running out. To give one example, according to IPCC (1992), the rate of deforestation in tropics is projected to increase in the future and 73% of all tropical forests are expected to be cleared by 2100. Further, climate change to which the world has committed may also have an adverse impact on BD. Surely it is possible to minimize the damage to BD through technological, financial and institutional arrangements and committed action at local, national and global level. Myers (1993) concludes, "The current mass extinction which is gathering force if proceeds unchecked, will not only eliminate half or more of all species, but will leave the biosphere impoverished for at least 5 million years – a period twenty times longer than humankind itself has been a species. Present society is effectively taking a decision on the unconsulted behalf of perhaps 100 trillions of our descendants, asserting that future generations can certainly manage with far less than a full planetary stock of species".

Acknowledgement:

This paper is a result of a number of studies conducted at the Centre for Ecological Sciences. I thank The Ministry of Environment for its support to the Centre.

References

Bedigian, D. 1991. Genetic diversity of traditional Sesame cultivars and cultural diversity in Sudan. In *Biodiversity*, Eds. Oldfield, M.L. and Alcorn, J.B. Boulder: Westview Press.

Clay, J. 1991. Cultural survival and conservation: lessons from the past twenty years. In *Biodiversity*, Eds. Oldfield, M.L. and Alcorn, J.B. Boulder: Westview Press.

Dugan, P.J. (ed.) 1990. Wetland conservation: A review of current issues and required action. Ghana IUCN.

FAO 1993. Forest resources assessment 1990; Tropical countries, FAO, Rome.

Gadgil, M. 1991. Conserving India's biodiversity: the societal context. *Evolutionary Trends in Plants* **5**: 3–8.

Gadgil, M. 1992. Conserving biodiversity as if people matter: A case study from India. *Ambio* **21**: 266–270.

Gadgil, M., Berkes, F. and Folke, C. 1993. Indigenous knowledge for biodiversity conservation. *Ambio* **22**: 1993.

Hobbelink, H. 1991. *Biotechnology and the future of world agriculture*. London: Zed books.

IPCC 1992. Climate Change: 1992, Supplementary report to IPCC Scientific Assessment, Inter governmental Panel for Climate Change, Cambridge University Press.

Malhotra, K.C., Debal Deb, Dutta, M., Vasulu, T.S., Yadav, G. and Adhikari, M. 1990. *Role of Non – timber forest produce in village economy*. Calcutta: IBRAD.

Myers, N. 1990. The biodiversity challenge: Expanded hotspots analysis. *The Environmentalist* **10**, 243–256. Myers, N. 1993. Questions of mass extinction. *Biodiversity Conservation* **2**: 2–17.

Myers, N. 1993. Biodiversity and the precautionary principle, *Ambio* **22**. No. 2–3, 74–79.

Ravindranath, N.H. and Hall, D.O. 1995. *Biomass energy and environment; a developing country perspective from India*. Oxford: Oxford University Press.

Ravindranath, N.H. and Somashekar, H.I. 1991. Biomass production, utilization and conservation in semi-arid area. In *Karnataka State of Environment Report* **5**, ed. Saldhana, C.J. Centre for Taxonomic Studies, Bangalore.

Reid, V.W. 1992. How many species will there be. *In Tropical deforestation and species extinction*, Eds. Whitmore, T.C. and Sayer, J.A. London: Chapman and Hall.
Ryan 1992. Life support: Conserving Biological Worldwatch paper 108. Washington.
Toledo, V.M. 1991 Patzuaro's lesson: nature, production and culture in an indigenous region of Mexico, In *Biodiversity* Eds. Oldfield and Alcorn. Boulder: Westview Press.
TSL (1992) Statistical outline of India, Tata Services Limited, Bombay.
UNEP (1995) Global Biodiversity Assessment, UNEP, Nairobi (in print).
WCMC (World Conservation Monitoring Centre) 1992. *Global Biodiversity*, World Conservation Monitoring Centre, Chapman and Hall.
WRI 1994. World Resources 1994-95, New York: Oxford University Press.
WRI-IUCN-UNEP 1992. Global Biodiversity Strategy, World Resource Institute, Washington.
Zuidema, G., G.J. van den Born, J. Alcamo and G.J.J. Kreileman 1994. Simulating changes in global land cover as affected by economic and climatic factors. Water, Air and Soil Pollution 76. 163–198.

10. Human Genetics: Liberté, Egalité, Hérédité

Philippe Marliere* and Rupert Mutzel**

* *Department des Biotechnologies, Institut Pasteur, 25, rue du. Dr. Roux, 75015 Paris, France.*
** *Fakultat fur Biologie, Universitat Konstanz, 78464 Konstanz, Germany.*

TOWARDS A STRANGE NEW WORLD

Mankind had a secret dream: Liberté, Egalité, Fraternité. For centuries, the eventual arrival of individual liberty, equality of rights, and solidarity appeared halted by political tyranny, racial discrimination and colonialism. With the decline of the communist empire the route appeared wide open to a "new world order". Humanity seemed to have reached the "end of history", and what now threatens our happy planet and its inhabitants (let aside local accidents in Somalia, Rwanda or Yugoslavia) are global menaces: overpopulation, the greenhouse effect, ozone depletion, BSE, trafficking of plutonium, or catastrophes arriving from outer space.

The clergy that promises to deliver us from these global evils is no longer priests or prophets, philosophers or politicians, but scientists, and its Scriptures appear weekly in *Nature, Science,* and diverse *Proceedings.* The past has clearly shown that the application of scientific knowledge can physically change the face of the earth-think of nuclear energy and there is no doubt that the application of scientific knowledge can be criminal. Science is always knowledge and the way to it. Like any human activity, the acquisition of knowledge can be criminal. In the case of human genetics, knowledge in itself will be criminal as soon as it refers to individuals and not to the "genus" Man. To put it in a pointed way: The determination of the entire DNA sequence of the genome of any human being (no matter whether black, brown, red, yellow, or white) is a very sober scientific task. Analysis of any piece of DNA from another individual for comparison will be criminal since it will immediately touch the individual liberty, equity, and solidarity that his peers owe to him and vice versa.

Several great minds have shown themselves alarmed by the dangers that could threaten mankind by the application of the spectacular progress in biosciences, pointing out above all the possibility of genetic manipulation of our descendance, either by positive selection of desired hereditary traits, or by elimination of "bad" genes from the population. Put under quarantine after the Nazi atrocities, euthanasia and the idea of "lebensunwertes Leben" (life that is not worth being lived) have seen a renaissance in the eighties, and there is a wide consensus that "Mengelian genetics" have to be outlawed from their very beginning. It is not within the scope of this essay to multiply these concerns on "constructive" molecular genetics, or to shout for more legal regulations. We will show that the analytical power of human molecular genetics alone has already started to shake our conception of man and the way societies deal with individuals. Progress in the biosciences thereby meets with latent and overt tendencies in medicine, social security systems, the judiciary, the labour market, and family and reproduction policy that result in the transformation of human subjects into objects of analysis and manipulation. This may lead to the loss of the individual's status as a free, equal, and responsible member of society, and hence to the loss of the individual's fundamental rights. Characteristic of this continental drift in the relation of the individual to social authorities is the transformation from a caring, curative attitude, via statistical, sociological preventive measures to the prediction of the individual fate and therefore to preventive treatment at the individual level.

MEDICINE GOES SCIENCE

Modern medicine is pushing forward most actively the disfigurement of human beings from subjects to objects. According to the protagonists of "predictive medicine" each of us carries a more or less heavy genetic burden predisposing us to freckles or colorectal cancer, hiccups or diabetes, schizophrenia or depression, cigarette smoking or compulsory gambling. First, this makes the frontier between endogenous pathologies and exogenously caused diseases to become blurred, and second, reinforces the cynical physicians' joke according to which there are only sick people and *healthy imaginaries.*

The main task and the declared aim of predictive medicine are so far analytical: to map and analyse at the molecular level these predispositions in order to establish, for each of us, a genetic file. Predictive medicine thus is the natural successor of preventive medicine which, during the last decades, has almost replaced curative medicine: the Hippocratic ideal of curing even without knowing is being perverted to a scientific idol, the compulsion to know even without curing. Medicine abandons its very human greatness in favour of getting the touch of Science.

Quite obviously, however, as the initiates of this new scientific discipline tell us, predictive medicine promises much progress both for the individual and for the entire society: knowing our risks may allow us to change our lifestyle and keep to a set of adequate rules in order to minimise the chances of explosion of genetic time bombs ticking in our cells, at the same time minimising the prospective social costs of diseases that cannot be foreseen at present. Moreover, keeping to these rules would promise both a longer life span and less anxiety for the individual. None of these arguments survives analysis. Our statistic life span has indeed increased very considerably during the last decades, owing at least in part to progress in medicine. In parallel, however, the incidence of complaints of old age has increased in the industrialised part of the world to a point where these diseases now cause a very major part of the social costs for health care. Eventually we will all die at the expense of society and no predictive policy will be able to avoid this fate and its associated costs. Even worse multiplication and wide application of predictive (and preventive) medical technology will increase the overall costs of health care, and reveal itself most profitable for pharmaceutical industry and the entire sanitaro-industrial complex. From the individual, on the other hand, predictive medicine demands no less than to abandon his fundamental democratic rights and duty, liberty, equality, and solidarity. We will be left with the liberty of choice between the knowledge of incurable menaces and a preventive medical treatment that can be escaped only at one's own risk, the equity of citizens marching in step towards normalisation, the solidarity of sacrifying our own fates on the altar of cost minimisation. Even if we let aside these dark prospects, we ought to doubt whether predictive medicine would better the quality of life in a society consisting of hypochondriacs, body-builders, and weight-watchers. Well before the advent of molecular genetics, western societies have seen the installation of an industrial health care which already deprives the patient of a significant part of his dignity and human rights. The pace of this trend will accelerate outside of the megaclinics and heart centres, putting physicians into conflict between their Hippocratic duty to get informed about the medical status of an individual patient and the vertiginous temptation of genetic archiving of the entire population.

PROFITEERS OF THE PREDICTABLE

At the same time social security systems, in the first place health insurance, will be faced with a dilemma. Indeed, their *raison d'être,* the administration of the non-predictable, is totally incompatible with the prediction of the individual fate. Private insurance companies have for a long time sold graded contracts according to age, sex, and the medical status of their clients (let aside their mental health). In other words, they have contemplated a sort of unconscious or unintentional genetic selection. In spite of the voluntary character of private health insurance this practice is fundamentally discriminating: imagine the public scandal that a night club would provoke, if it refused entry to a disabled person arguing that this could increase its running costs. By contrast, there is the very real case of US health insurance companies refusing to accept children of couples at risk of a genetic disease unless prenatal genetic testing is performed. If the test would reveal a genetic defect: would the parents find any health insurance for their baby? It is a most remarkable fact that this happens in a society where, at the same time, "abortion doctors" are shot, and where political correctness urges us to call a disabled person "differently abled".

Discrimination will multiply with increasing power and resolution of genetic diagnoses. Paradoxically it will most probably enter the system by proposing discount contracts for genetically well-off people. In the end, we shall see the agony of the idea of a social security system based on solidarity among the individuals who constitute it. Instead, the charges will be put on each individual according to his presumptive genetic fate. Once again, our individual financial means will define our health.

MANAGING HUMAN RESOURCES

Just as a distorting mirror cannot invent reality, genetic exploitation and its brutal consequences do not result from molecular biology itself; they rather follow pre-set avenues towards domestication and adaptation of human beings by their own kind. Employment strategies are paradigmatic for this who would enter an airplane without being convinced that members of the crew have been selected for their good eyesight, perfect reflexes, and reliability? At a first glance, this evaluation of human aptitudes as the products of inborn qualities rather than a professional education appears to be a matter of course; there is rarely any reflection on this selection even by the candidates who are being evaluated. In fact, this matter of course hides a technological decision arrived at by weighing the human material against a host of other material factors, the human factor thus becoming one of many in a cost-to-benefit calculation. From the passenger's point of view, a thoroughly selected captain stands as the positive aspect of "genetic" prediction.

Human genetics, however, promises more than that, namely the possibility to pass from the potentials of persons to potential persons. Again, the field is well-prepared. The mere fact that employers attempt to predict the potentials of their employees by physiological, psychological, sometimes even graphological and astrological parameters in addition to their documented education record demonstrates that there is a vast market for the fully transparent potential person which extensive analysis of genetic characters and their prospective future "phenotypic" manifestations will provide for. By then only will it be possible to thoroughly manage human resources, and to adapt these resources to the conditions made dirty by ourselves; consider the enormous savings that chemical or nuclear industries could make if the solvent or radiation resistance of each worker could be reliably predicted, allowing to economise, as one of many cost factors, protective measures. As a consequence, employers will no longer have to care for their personnel or *prevent* them from damage, but merely predict their potential.

JUST SCIENCE

Justice has an even longer history and experience in evaluating and judging Man, and justice is drifting away from its former curative duties – atonement and punishment for committed crimes – via sociological prevention to the prediction of crimes and therefore to preventive measures at the level of individual persons. Molecular genetic techniques are now used routinely as forensic means for identifying, at a presumed very high level of confidence, if not unambiguously, capital offenders; genetic "fingerprints", a sort of specific bar code for every individual (except for identical twins) are presented as strong evidence in murder and rape trials. These methods still demand to be performed and supervised by skilled scientists, and they remain quite expensive. It can, however, easily be foreseen that concomitantly to miniaturisation and automatisation of the protocols, and digitalisation of the results and interpretations, the market will behave in a similar way as previously the market for computer chips behaved. It will thus become straightforward to apply molecular genetic forensics in trials of far less capital character, for the prosecution of ever lighter offences, and the surveillance of subversives, radicals, or "habitual criminals". There is probably no modern police in the world that does not at least plan to establish a database containing the genetic profiles of convicted criminals. Moreover, most of these databases will be systematically filled with data of suspects that later will prove innocent. Police officials frequently concede that it may be "difficult" to remove these data later. In addition to their use for classical forensic purposes, such databases will sooner or later be open to other administrative and executive authorities – there is already the precedent of immigration officials in Britain demanding genetic proof of the identity of family members who want to join a relative resident in Britain.

The dangers of these projects and practice for the status of individuals in democratic societies have so far provoked only very limited protest, probably owing to their technical complexity and incomprehensibility for the layman. Moreover: they concern criminals and immigrants, so what should we fear if we have nothing to hide? Until now, the main criticism was raised by molecular biologists themselves who, apparently less confident about their own technology than the judiciary are stressing that genetic testing has to be performed in a strictly controlled, scientific manner in order to avoid any miscarriage of justice. However, their concern was mainly that forensic laboratories could produce flawed results, and virtually none of these experts ever shuddered at molecular inquisition per se which may today judge on our guilt via the analysis of a single hair or drop of blood, and which may judge tomorrow our criminal potentials.

For a long time, justice has been searching the Holy Grail of the transparency of the criminal, confession; this was evident well before the masterworks of modern biology entered the courtroom. Remember the tragicomic lie-detector seances in the United States. Just like today's molecular genetics, the lie-detector was criticised precisely by its own inventors for its technical unreliability. Neither its technological closeness to the electric chair (similar to our *healthy imaginaire*, a defendant is a guilty person just not yet convinced), nor the ridiculousness of a democratic system that from time to time subjects even its ministers to a prophylactic lie-detector test, and that even saw a president trembling with this creator of transparency, raised any criticism. While the lie-detector promised – by a very complicated procedure – the transparency of individuals in a few rare cases, molecular human genetics will lift off the opacity of the entire population. To give just one example for possible applications, imagine the ease with which the biological father of every child in every family could be determined, and imagine the social consequences if the data would be revealed. After all, this example is still quite harmless compared with the sum of the consequences of systematic genetic analyses on the fundamental rights of each citizen.

Human rights are based on the concept of individuality, and this concept will eventually vanish. Criminal laws make a discrimination between the responsible offender committing a crime fully conscious to act unlawfully, and a person that is not responsible for his action at the moment of the crime. We now have to be prepared for trials populated by witnesses that are all genetically more or less predisposed for lying, with experts according some confidence level to each of them.

Our present conception of free will, perhaps modified by mitigating or aggravating circumstances would mutate to the scientific definition of mitigating or aggravating personalities, uniting Kafka's Accused with our *healthy imaginaire.*

A PACKAGE TOUR TO PARADISE

Quite obviously, this deformation of democratic society is possible without any manipulation of our descendance, without any eugenics. However, it is a small step from here to eugenics – just apply the arguments of predictive medicine on future generations to see the phobias and fantasies of the consumer society become true in actual persons. Prenatal selection or, initially, counterselection, of hereditary predispositions has the advantage of avoiding to make "real" victims; *therapeutic* abortion can well be founded both on eugenic and prophylactic reasons. Breeding a happily domesticated human race will therefore at no defined stage really touch the "free will" of individuals, even if we are faced, during the course of the process of handicap elimination, with certain exaggerations like those we see in the sports business today.

A planned genetic economy defined by molecular physicians, gene technocrats, employers, and insurance companies will terminate our existence as individuals even before it begins. We will be the passengers of a package tour to a predetermined destination, the best of all worlds, deprived of any means to change its direction at any point. Our stay in the genetic Club Mediterrané will be animated by benefactors of humanity fighting against disease and handicap. No doubt the public will be satisfied, and even ask for more: we are ready to step from the by then emancipatory maxim "a child when/if I want it" to "a child how I want it", from birth control to democratic eugenics. In contrast to genocracy we can find a positive aspect in nuclear society: it was easily seen through since it required a repressive apparatus to install itself. We cannot expect, if not hope for, a genetic Chernobyl that would awake the public opinion.

The criticism that is sporadically raised against these developments appears well-inspired but badly advised. The "ecological" warning "don't touch the genetic diversity of mankind, it's evolutionary potential", is dangerous for two reasons. First, it asserts that we know what is of advantageous for the future of mankind. Second, we can easily think of eugenics aiming at maximising certain polymorphisms by using the same argument, and leaving selection of the actual genetic best-seller to the public. Another argument, more generous but also more dangerous, says that genetic determinants are of very minor influence on "real" people. We have already seen, on the pages of *Nature* or *Science,* family trees "demonstrating" hereditary transmission of violence, alcoholism and worse. It is a matter of a few years to see similar trees describing the hereditary propagation and molecular nature of the mutations causing aberrations, such as addictive gambling, under-employment, or nail-biting. The reaction of our societies to this sort of scientific progress will be the litmus test for their humanism: it is our distinguished human duty to fight against such determinism, and to prevent the analysis and mapping of its underlying mechanisms in the individual, *even if it exists.*

It is illusory to think of legal barriers halting these developments. Special regulations are subject to change – think of embryo research in the United States, or of regulations on genetic engineering in the European Community which appear to reflect rather the state of the art than any moral limitation to future applications of the art. Laws against genetic discrimination should rather be considered as a manifestation of the current common will to avoid the current "worst", and they may serve to cut off the tip of the iceberg. India, for example, has recently passed a law that reveals a very noble-mindedness in the leading political class, we doubt that it will save the lives of many female foetuses, let alone newborn or adolescent girls. There is probably no barrier to such practice unless it is fundamentally outlawed by the entire society, that is, unless it overthrows taboos.

Apparently no scientist has ever tried (or, at least published) the creation of a human-chimpanzee chimera (although mice carrying elements of the human immune system are already in use). Such a chimera could be of enormous medical interest, and it should be easily feasible from a technical point of view. It is not being made precisely because it would violate a taboo. The central question is therefore whether society can create taboos halting the previously unthinkable. We should be pessimistic about the answer. We should also be cautious about the fundamental noble-minded scientists, physicians, and jurists, not to speak of employers and insurers. The demonstration that Dr. Mengele and other Nazi scientists were true psychopaths is still lacking, and their "scientific" data now happen to be cited as regular reference. Moreover, society should carefully take note of the leading articles in the scientific press and insist on its power to decide whether public money should better be spent for the amelioration of the life conditions of the underprivileged, or for analyzing whether they carry gene predisposing them to be underprivileged.

In this essay we tried to pinpoint some of the dangers which threaten the foundations of democracy by the systematic use of human genetics. These dangers do not result from the logic or the applications of modern biology; they are rather the consequence of its transfer to sensitive areas of society. Medicine, insurance companies, and employers have long been contemplating rudimentary, statistical, possibly even unconscious genetic analyses, that threatened liberty, equality and solidarity between human beings. Now that these practices can be formalized and scientifically justified, they make the nature of man appear incompatible with the democratic ideal as it was formulated in the eighteenth century. Even if we are not Machiavellian we have to consider that human genetics may be pushed forward under a shield of incomprehensible scientific high-tech, just as the growth of the "miliary-industrial complex" has been possible under the shield of the incomprehensible complexity of modern weapons. There is definitely a way out of this dilemma: school systems have shown that we are well-armed against any form of determinism, as long as fundamental equity and deliberate acceptance of individual opacity are estimated higher than technological feasibility and economical constraints. The education system in most societies has strongly refused systematic IQ testing as a predictive means to determine the fate of its charges, pointing to their principal and fundamental equity of chances, and measuring them by their progress and actual performance instead of their "nature".

L'HOMME EST-IL FAIT POUR LE BONHEUR?

Probably for the first time in human history we can and must address this cruel question. The ideals of democracy were born in the drive of rebellions and revolutions, forged in the fire of oppression. The last great menace that western democracies had to fight against was Nazi barbary. There are no more dangers to them from outside, therefore the definition of the principles that will guarantee the survival of democracy have to arise from the very heart of democratic society. While we have established rules that assure good relations between diverse social groups, detailed formalization of the relations between individuals and collective instances have been left blank for good reasons. Society now has the choice of either continuing to respect each individual together with its weaknesses and handicaps, to ignore it genetically, or to scientifically measure our values and potentials and to allocate us our place for the sake of common wealth and social economy, in other words, to allow the re-instalment of "natural selection" in human society, or even, to actively reinforce it.

Index